Site Interaction and Political Geography in the Upper Usumacinta Region during the Late Classic: A GIS Approach

Armando Anaya Hernández

BAR International Series 994
2001

Published in 2019 by
BAR Publishing, Oxford

BAR International Series 994

Site Interaction and Political Geography in the Upper Usumacinta Region during the Late Classic

ISBN 9781841711959 paperback
ISBN 9781407353470 e-book

DOI https://doi.org/10.30861/9781841711959

A catalogue record for this book is available from the British Library

This book is available at www.barpublishing.com

BAR Publishing is the trading name of British Archaeological Reports (Oxford) Ltd. British Archaeological Reports was first incorporated in 1974 to publish the BAR Series, International and British. In 1992 Hadrian Books Ltd became part of the BAR group. This volume was originally published by John and Erica Hedges in conjunction with British Archaeological Reports (Oxford) Ltd / Hadrian Books Ltd, the Series principal publisher, in 2001. This present volume is published by BAR Publishing, 2019.

BAR
PUBLISHING

BAR titles are available from:

BAR Publishing
122 Banbury Rd, Oxford, OX2 7BP, UK
EMAIL info@barpublishing.com
PHONE +44 (0)1865 310431
FAX +44 (0)1865 316916
www.barpublishing.com

ABSTRACT

The reconstruction of the political organisation of the Maya lowlands during the Classic period has been the subject of continuous testing and debating of explanatory models. These have ranged from the strong centralised state of the Old Empire to the ephemeral peer polity/fragmented states models. Except for Puleston (1974) who argued on behalf of approaching a regional study from the periphery, everyone else has focused his or her studies from the centres. Hence most archaeologists have readily adopted models derived from Central Place Theory and Nearest Neighbour Analysis. However in order to apply these models archaeologists had to draw the line between primary centres and subsidiary centres.

Before the advent of the "epigraphic revolution", site hierarchy was determined by quantitative methods, i.e. number of monuments present, number of courtyards, volumetric analysis, etc. With the discovery of the nature of the emblem glyph the pursuit of a site hierarchy continued, but following a different path. Based on their presence or absence the emblem glyphs where used to differentiate primary centres from subsidiary centres. But what about hierarchy between those sites that did have an emblem glyph.

In this context first Barthel (1968) and later Marcus (1973) argued for the existence of four regional capitals in the Maya world. Mathews (1991) after a more detailed analysis of emblem glyphs concluded that these denoted the existence of completely autonomous polities. Hence the debate continued between those who favoured a strong centralised state, and those who favoured a scheme of basically small autonomous states.

Recently Grube and Martin (1995, 1998) have come up with a "middle ground" model. Through their analysis of a series of glyphs that express political subordination they have proposed the existence of two hegemonic powers, Tikal and Calakmul. In this scheme virtually the rest of the Maya world revolved in a succession of political alliances. Grube and Martin's discovery seems to have resolved the contradictions that the two previous models were unable to resolve.

Aliphat (1996) in his doctoral dissertation reached the conclusion that the physical landscape of the Upper Usumacinta played a pivotal role in shaping the settlement pattern. He also advanced the idea that access to, and control of the principal communication routes was a priority to the kings of the region. I followed this line of research in the present thesis, arguing that it was precisely this determinant factor in regional supremacy.

In this work I apply a conjunctive approach drawing from two lines of research to offer a reconstruction of the political organisation of the Upper Usumacinta region: Geographic Information Systems (GIS) and the hieroglyphic inscriptions. In the first instance I begun by creating a Digital Terrain Model for the region in order to model movement across the landscape. This enabled me to reconstruct territorial size and communication routes between polities based on cost-surface analysis. Later, I used an approach similar to that used by Tobler and Wineburg (1977) who applied the Gravity Model as a predictive model for the location of a series of sites in pre-Hittite Anatolia. The Gravity Model works on the premise that interaction between centres will be more intense if these are located closer together, and will decrease as distance increases. The inscriptions contained in the study contain mention of various presently unallocated sites. Of these Sak Tz'i', Hix Witz, Man, and the "Knot-Site" are mentioned by more than one of the regional centres.

I used these references and quantified the degree of site interaction in order as a measure of distance that would help me locate the unidentified sites. The Gravity Model identified the most likely areas for the location of Sak Tz'i', Hix Witz and the "Knot-Site", the distances obtained for Man set this site outside the study region.

The Dempster-Shafer logic, which was used in the construction of is a probability model, was also applied to the data set. The results of Dempster-Shafer support those obtained with the Gravity Model. With these results and the site interaction information available for the region I proceeded to create a series of maps that offer a reconstruction of the political organisation of the region. Instead of presenting these maps based on fixed time periods, the critical points in the history of the region were identified to obtain those maps. These results offer a more dynamic view on how the political contestants of the Upper Usumacinta shifted their alliances from one to another of the hegemonic powers, and how these changes were reflected in the political geography of the region.

ACKNOWLEDGMENTS

Many people and institutions were directly involved in the research and subsequent writing of this work, family, friends, and colleagues were a constant source of inspiration and knowledge. First of all my sincere appreciation to my dear friend and mentor Peter Mathews; in Peter I have found not only an inexhaustible source of knowledge; I have also found refuge in my many times of despair. Jane and Dave Kelley have also been critical in the development of my academic and personal life, both have opened their home and their hearts to me making the longing for home much more bearable. Scott Raymond went well beyond his duties as the Head of the Archaeology Department at the University of Calgary and has always been there for me, to him my deepest gratitude.

If I were asked to single out which were the turning points in my professional career I would refer to two persons. One I have already mentioned Peter Mathews. The other is Nigel Waters from the Geography Department at Calgary. Nigel introduced me to an exciting dimension of spatial analysis. He patiently walked me through the otherwise cumbersome paths of spatial analytical methods and GIS that had always eluded me. Stanley Guenter has also made a decisive impact in the development of this work. Like a yapping puppy he would storm into my office sharing with me his new and exciting discoveries in the decipherment of a Maya inscription. This led to long discussion sessions, which gradually gave shape to my Chapter Five. Thanks are also due to Dr. Duane Marble, who kindly provided me with an exhaustive bibliography on movement over natural terrain, and his own paper on the same topic. These proved to be invaluable for the construction of my working model.

In Mexico two persons also represent landmark moments in my professional life, Roberto García Moll and Mario Aliphat Fernández. Roberto drew me into the Maya area trusting me with his project at Bonampak. Mario honoured me with his friendship during the good times and especially during the bad times. It was then when he invited me to participate in the Proyecto Arqueológico El Cayo, and that in turn opened the doors to my postgraduate studies at Calgary. My recognition to my Mexican colleagues, Alejandro Tovalín, a superb person and archaeologist, who unselfishly has shared much of his data from Bonampak with me and has provided his generous hospitality during my visits to his camp. By the same token Arnoldo González, Daniel Juárez and Akira Kaneko have always given me proof of their generosity.

On the institutional level my gratitude goes firstly to Mexican Council of Science and Technology CONACYT, which supported me through my studies at the University of Calgary. In Canada I also received the generous support of the Archaeology Department, the McMahon Stadium Society and the Dean of Graduate Studies Office. Finally in Australia I have received the overwhelming support of La Trobe University through the generous intervention of Professor Tim Murray. Without these institutions I could not have accomplished this work.

I leave my family for the last because they fall in a completely different category. The level of their support is of course immeasurable. My lovely wife Yanira has been the driving force behind me throughout these years. She has suffered in silence the rigors that the life of a postgraduate student imposes. To my children Axel, Ixchel, Raul, and Max my love and my apologies for not being there all of the time. To my parents my endless love and admiration; throughout my life they have always encouraged and supported me in chasing my dreams. I thank my brothers Adolfo and Mario and my sister Leticia who through their example have shown me the value of family bonds; they have always been supportive and have made my existence more cheerful with their joyous view of life.

DEDICATION

To the three women of my life: my mother, my daughter, and my wife, without you I would be lost.

TABLE OF CONTENTS

LIST OF TABLES

LIST OF FIGURES

LIST OF GRAPHS

ONE SPACE, THE FINAL FRONTIER

Although every archaeological study, past and present, has some spatial component, nevertheless the archaeological discovery and conquest of space has only recently begun on a serious scale.

Introduction

Archaeology, since it left behind its initial endeavours of constructing a culture history to concentrate in the more challenging problem of explaining cultural processes, has been in a constant pursuit of finding the right formula to handle its data under the rigorous rules of the scientific method. To achieve this goal archaeologists have turned to other more mature disciplines borrowing their analytical methods and customising them to their own needs. In this incessant search of the Holy Grail we have ventured into the perils of the uncharted fields (for most of us) of the "hard" sciences in an attempt to grab as much as we can and recycle it to our own advantage. Frequently, however, like the Maya hero twins that descended to *Xibalbá*, the gods of this epistemological underworld tease and torment us. We think we have the key to break the code that will make our data speak their truth, only to find that we deceived ourselves, that this wasn't quite the right formula.

During the early 1970's, at the peak of the momentum that the New Archaeology had created, some archaeologists believed that they finally saw the light at the end of the tunnel, and thought that archaeology had come of age in the scientific method. In 1971, a very important event took place: the *crème* of our discipline gathered together to hold the seminar *The Explanation of Cultural Change: Models in Prehistory.* The title of the seminar was impressive enough to warrant the belief that something big was taking place. One after another, archaeologists of this new paradigm presented the fruits of their arduous research, which in effect addressed the issue of explaining cultural change. Invited to make the closing commentaries was a scholar of enormous stature, the prestigious Edmund Leach. In an exercise of good will and confidence towards their newly acquired analytical tools the organisers decided to invite this renowned scholar of a related discipline. I don't think that what the participants of this seminar heard was precisely what they expected (or wanted). Leach addressed the audience saying that during the past few days he had seen the presentation of one model of cultural change after another. His characterisation of these models is one of my favourites because it captures the main flaw of what an explanatory model can become. Leach told his apprehensive audience that their models reminded him of the aeroplane models that one can buy in a toy store. There are two kinds: those that look like the real thing but won't fly, and those that do not look like the real thing but can fly (Leach 1973:761-771).

Leach continued decrying the abuse in the application of these models for issues that are out of the reach of archaeology. He made special emphasis on the limited character of the archaeological data. Next he cautioned archaeologists that when they start moving from questions regarding the formal nature of their data to questions such as how and why this came to be, they were moving from the realm of the verifiable to that of speculation (Leach 1973: 764).

Other cultural anthropologists echoed Leach's critique and pointed their guns at some of the more ambitious models that were flying around. See for example the rather vicious critique that Allen and Richardson made to Longacre's, Hill's, Whallon's, Binford's, and Deetz' attempts to reconstruct prehistoric kinship systems from their archaeological data (Allen and Richardson 1971:41-53).

Are we to assume then that archaeology is doomed to never get a stronger grip in interpreting its data? Are models inherently flawed and their application to solve the firing issues of archaeology bound to failure? To answer these burning questions I recur to yet another definition of model. This one I got from Clarence Woudsma, faculty member of the Geography Department at the University of Calgary. Dr. Woudsma compared models to sausages: "everybody likes them until they know what they're made of". This colloquial and apparently simple characterisation of what a model is reveals a great truth. What we know or don't know of a model is going to greatly affect the results of our analysis. If we put our data in a computer, press some buttons to start running it but we don't know what's taking place inside the black box, we will be ill equipped to fully interpret the results.

In this day and age of computer technology Geographic Information Systems (GIS) are rapidly gaining acceptance in archaeology as a powerful analytical tool. GIS is being hailed in virtually every discipline as the answer to problems of data management and manipulation. To some it has become the panacea; others take it with more reservations. Whatever the case the fact remains that it's out there and we have to pay attention to it.

For the problem that concerns me in this work --the reconstruction of the political geography of the Upper Usumacinta during the Late Classic--I believe that GIS represents an asset in the interpretation of the available data. Through the application of GIS to my research data, I believe that I have gained further insight in the way these is interrelated, especially when the sheer bulk of it prevents me from "eyeballing" through it to gather relevant information. I'm not concerned with the fact that with the adoption of GIS as an analytical tool I am following suit with the tradition in archaeology of borrowing from another discipline. The spatial nature of our data allows it. However I am also aware of the temptations that GIS may offer and have tried to use extreme caution in order to avoid putting the cart in front of the horse, so to speak.

The research problem advanced above is framed within the context of Landscape Archaeology. In this sense I follow and build on Aliphat's 1994 doctoral dissertation. In his results Aliphat concluded that the characteristics of the physical setting of the Upper Usumacinta Valley greatly influenced the location of the more important archaeological sites, and that the location of these sites obeyed what Hammond (1975) defined as the strategic factors, which in this case are related to their proximity to the critical points along the main communication routes. To reach these conclusions Aliphat thoroughly described the natural landscape. This description was based for the most part on his own empirical observations in the field.

Aliphat however had to limit himself to addressing the nature of his data without being able to raise *what if?* type questions. In this work this is one of the venues that will be followed. In order to propose a reconstruction of the political geography of the region I need to elucidate which were the relevant centres and where and how they were placed in the physical and socio-cultural landscape.

The hieroglyphic corpus of the region describes an intense interaction between the sites located in the Upper Usumacinta region. We know the location of most of these sites, but there are a few that are frequently mentioned and whose location is still unknown. On various occasions Peter Mathews had mentioned to me the plausibility that one of those sites, Sak Tz'i' would be found in the vicinity of Laguna Santa Clara, on the Mexican side of the Usumacinta River. Likewise for Hix Witz -another of those unidentified sites- he suggested that it may be located between Piedras Negras and Yaxchilan, on the Guatemalan side of the Usumacinta. Knowing of the keen reputation that Peter has in closing in on unknown sites (e.g. Site Q) I paid attention to what he said and decided to pursue the matter further. If I could determine the whereabouts of these sites (key participants in the political scenario of the region), then I would be in a better position to understand the mechanics behind the strategic factors proposed by Aliphat. This in turn would help me to model the waxing and waning of the different regional polities and the possible extra-regional influences that shaped the political map of the Upper Usumacinta region.

Through the joint application of GIS and the epigraphic evidence I believe that I'm in a position to attempt such an endeavour. However in order to lend credibility to the results obtained, I must take very seriously Dr. Woudsma's sausage analogy. In this sense this work has been organised in the following sections.

Chapter One deals with the theoretical underpinnings of spatial modelling. Here after a short history of the role that geography has played in the theoretical development of archaeology I review the three most frequently used spatial models in archaeology: Central Place Theory, Nearest Neighbour Analysis and the Gravity Model.

In Chapter Two I review the development of settlement pattern studies for the Maya lowlands. I present some examples of how the above listed spatial models have been applied in the interpretation of site distribution and conclude by presenting the contributions that the epigraphic interpretations have given to this area of archaeological studies.

Chapter Three contains the introduction to the physical region. The basic concepts of Landscape Archaeology are defined in order to show how the physical setting is conceived for spatial analysis. The role of the communication routes in relation to this physical landscape is emphasised in this chapter. To better illustrate this role I present the accounts of the earliest documented *entradas* into the region by the Spanish conquistadors, and those of the late nineteenth century traveller/archaeologists Alfred P. Maudslay, Desiré Charnay, and Teobert Maler.

Chapter Four is where I make explicit the contents of my working models (or to keep pushing the sausage analogy, the ingredients). I have divided this chapter into two parts. Part I contains an introduction to the fundamentals of GIS and its application in archaeology. In this section I go into some detail concerning the criteria used in constructing a Digital Terrain Model (DTM). The criteria established in the definition of the algorithms used for modelling movement over natural terrain, and conclude with the characterisation of predictive modelling for site location purposes in archaeology. Part II contains a detailed description of the methodology used in the spatial analysis. The parameters used in the construction of the DTM of the study area are defined, as well as the criteria used in creating the cost-surfaces that were used in modelling movement across the terrain and the definition of polity extension. In this section I also define the application of the Gravity Model and present the results in terms of the potential location of the territories of the unknown polities. Finally, I introduce Dempster-Shafer Theory as a site prediction model and compare the results obtained with those provided by the Gravity Model.

In Chapter Five I return to the socio-cultural dimension of the landscape and set the results of the above-described models in the framework of the epigraphic data available. Movement across natural terrain is once again dealt with, but this time from the perspective of the control of the main communication routes in the context of site interaction. A reconstruction of the political geography of the region through time is offered, but this time instead of presenting maps at regular time intervals these are presented on the basis of the crucial historic events that took place. A discussion of the extra-regional influences is included as well as a structural characterisation of site distribution.

Chapter Six contains the concluding remarks. Two issues are addressed, the first deals with the observed patterns in site location and levels of interaction. Here I attempt to characterise the nature of the settlement patterns of the Upper Usumacinta region and identify the forces behind this distribution. The second set of conclusions deals with the pertinence of the spatial models used for site prediction purposes, and in more general terms with the use of GIS as a heuristic tool in modelling the past.

There are three appendices included that have very specific information that, although relevant, are too lengthy to include in the body of the text:

Appendix 1 contains the cartographic model of the majority of the image production/analysis process. Minor editing details or repetitive stages were left out for the sake of space.

Appendix 2 contains the summary statistics for the preparation of the Gravity Model. The complete data matrix was too big so I did not include it.

Appendix 3 is a table containing all of the site interaction information that I had knowledge of and was included in the Gravity Model. It is organised chronologically, containing six fields: Long Count position, Julian date, Site, Event, Other site mentioned and Monument containing the inscription.

Spatial Analysis and Archaeology

Like geographers, archaeologists think spatially. We make sense out of our data by referring it to its spatial dimension. Archaeology –and particularly European archaeology- has been closely linked to geography throughout its theoretical development. It is perhaps then, no surprise that many archaeologists have found in the Royal Geographic Society and the National Geographic Society an important forum to present, discuss and promote their findings. Clarke in his seminal 1977 work *Spatial Archaeology*, gives a detailed account of the influence that geography has had in European archaeology. From the formal mapping methods of attributes and artefacts developed by the Austro-German school of "anthropo-geographers" in the 1880-1900's, to Fox's (1922) method of combining archaeological and environmental maps through time (Clarke 1977:2-5). However it is with the Cambridge School of New Geography, and in particular Haggett's 1965 *Locational Analysis in Human Geography*, Chorley and Haggett's 1967 *Models in Geography*, and Chisholm's 1968 *Rural Settlement and Land Use*, that the otherwise implied influence of geography to archaeology is formalised.

This paradigmatic shift can be witnessed in Vita-Finzi's and Higg's 1970 site catchment analysis of Mount Carmel, Hodder and Orton's 1976 *Spatial Models in Archaeology*, and Clarke's own 1977 *Spatial Archaeology*. North American archaeology, however, was not as deeply influenced by geography as European archaeology was. Phillip's et al. (1951) work in the lower Mississippi Valley, and Willey's Viru Valley (1953) project sparked the interest in the spatial distribution of artefacts and sites across the landscape. However, the academic tradition of North American archaeology –with a greater influence of anthropology in its functionalist phase- tipped these studies more towards the social rather than the spatial aspects (Clarke 1977:2-4). More recently geography is once more marking the pace of theoretical development in archaeology in the form of Landscape Archaeology, which stems from Landscape Theory (see for example Green 1990, Crumley and Marquardt 1990, and Aliphat 1994). Within this context the application of Geographic Information Systems (GIS) has been hailed as the new paradigm (Allen, Green and Zubrow 1990; Gaffney and Stancic 1991). Currently the works of geographers such as Marble (1990,1996), Decker (Stine and Decker1990), Harris (Harris and Lock 1990), Lanter (Stine and Lanter 1990) –just to mention but a few- are having a comparable influence as Chorley's Haggett's and Chisholm's did three decades ago. How much can the application of GIS assist us in the interpretation of the archaeological data? Can the reconstruction of the political geography of the Upper Usumacinta valley during the Late Classic Maya be aided through the application of GIS? Before going into the specifics let us review some of the most popularly accepted geographic models in archaeology.

Clarke was very emphatic when he declared that the analysis of the spatial relationships of artefact and site distribution could no longer be done by intuitive methods such as "eyeballing" or simple inspection (Clarke 1977:5). The recognition of the non-stochastic nature of this spatial distribution has for long now been the guiding principle of spatial archaeology, the study of settlement patterns and Landscape Archaeology (e.g. Willey 1953; Hodder and Orton 1976; Clarke 1977; Ashmore 1977; Aliphat 1994; Green 1990; Crumley and Marquardt 1990).

This non-randomness is expected due to the equally non-randomness of human behaviour which according to Hodder is construed and determined (Hodder 1977: 224). Hence archaeologists will attempt to derive from the analysis of the spatial distribution of their data the explanatory models that account for the human behaviour which produced that patterning. However, due to the inherent limitations of the archaeological record (e.g. incompleteness of the data due to differential preservation, or sampling strategies) the explanatory procedures are not as straight forward as we wished. To circumvent these limitations archaeology has turned to other disciplines to help in designing its own explanatory models. Clarke (1977) identified four general theories that according to him underlie those archaeological spatial analyses that go beyond the descriptive level, these are: anthropological spatial theory, economic spatial theory, social physics theory, and statistical mechanics theory (Clarke 1977:18-20).

Anthropological spatial theory has been traditionally used in archaeology as a means of explaining human behaviour. It is based on the assumption that human behaviour can be gauged by defining the spatial patterning of the archaeological remains by quantitative methods, in order to offer testable hypotheses regarding social organisation derived from ethnographic analogy (e.g. Longacre 1966; Hill 1966).

Economic spatial theory is based on the assumption that as a result of empirical knowledge people will tend to adopt choices and solutions which minimise costs and maximise profits (e.g. Marcus 1983; Ball and Taschek 1991; Inomata and Aoyama 1996).

Expectedly the social physics approach draws its analogies from the laws of physics (such as the Gas Laws, Newtonian Laws, etc.). The central tenet of social physics theory is that although individual human actions are unpredictable, the joint actions of a large number of individuals (a community) may form predictable empirical regularities. In this approach the researcher does not necessarily try to explain why a specific spatial pattern is formed, but merely observes it in an attempt to find an empirical regularity in the formation process, which will enable him to reconstruct or simulate its occurrence. An example of the use of a social physics approach in archaeology can be seen in the application of the gravity model for predicting interaction between sites and populations (e.g. Tobler and Wineburg 1971; Clarke 1972; Hodder 1974).

Statistical mechanical theory represents a link between the social physics approach with the logic of statistical inference and the Likelihood Law. Through this approach the system can be successfully described as a whole, without having to know or describe in detail each and every part that comprises it.

The relevance of these underlying theories in archaeology depends on the level of spatial integration that is being addressed. Clarke identified three basic levels:

a) The micro-level, which analyses the spatial relationships within the structures, where proxemic and social models become more appropriate since individual and cultural factors prevail largely over economic factors.

b) The semi-micro level, spatial relationships are analysed within sites (between structures of the same site). This level comprises the communal space, where social and cultural factors may outweigh economic factors, and therefore social and architectural models are most appropriate.

c) The macro-level, yields a more regional perspective, relationships between sites are analysed at this level, where economic factors dominate most social and cultural factors, thus the adequate models for this level are derived from geography and economy (Clarke 1977:11-14).

It is the macro-level that is of relevance for the study of the settlement patterns, for which the most borrowed geographic-economic models are those for Central Place Theory, Nearest Neighbour Analysis, and the Gravity Model. I will describe briefly each of these in turn:

Central Place Theory

Most settlement pattern models in archaeology have been derived from geographic models originally developed for market economies, such as Von Thünen's (1826) and Weber's (1909) locational models, and Christaller's (1933), and Lösch's (1941) Central Place Theory. Von Thünen's recognises the relationship between the spatial distribution of activities and land use around a centre based on the "law of diminishing returns with distance". Thus the effect of this law will produce a series of concentric circles of land use and activity patterns around the isolated centres which reflect a behavioural tendency to maximise the returns for the least effort (Clarke 1977:21). In archaeology Von Thünen's model is the basis of Vita Finzi and Higg's (1970) Site Catchment Studies. The shortcomings of this model, however, lie in the basic assumptions of the uniformity of the terrain and the isolated nature of the centres.

Weber complemented this model by considering the location of the centre in terms of its external connections and the movement of resources from one place to another. In this context Weber's main proposition is based on that site location which minimises unnecessary movement. In other words site location with respect to resources was conceived in terms of minimum-energy least-cost. Weber's model applied with the appropriate precautions, has evidently a great potential to address issues such as specialisation and/or organisation of craft production both at the within-site level, and at the regional level when location of the sources of raw materials is taken into consideration (Clarke 1977:22-23).

Christaller successfully integrated the two previous models, taking into account the relationships between the area served by sites, site function, and the networks created between sites. He first considered all the sites clustered in an undifferentiated landscape, and arranged them in a hierarchical structure based on the type of resources and services that they provided each other. In this context in order to define the optimal least-cost organisational structure within the site network, Christaller analysed the demand of goods, resources, and services in relation to distance from distribution sites (Clarke 1977:23).

Central places provide a greater variety of goods and services than the smaller sites. In this sense central places are structured in a hierarchy comprising discrete groups of centres, in which centres of each higher order group perform all of the functions of the minor centres plus a set of essential functions that differentiates them from the lower order centres (Hodder and Orton 1976:60-61). Christaller proposed three basic patterns of hierarchical distribution of central places (Figure 1-1):

- The ***K* = *3*** system, where second level centres are arranged according to the market principle and are best situated for access to 3 larger centres.
- The ***K* = *4*** system, based on the transport principle where due to the importance of the communication lines that link primary centres, secondary centres are placed at midpoints between them.
- The ***K* = *7*** system, based on the administrative principle in which secondary centres are wholly encapsulated within the area of the larger centres.

Christaller showed that the sites in a specific network would tend to adopt a hexagonal territorial tessellation of space (Clarke 1977:24). In support of this discovery, Haggett (1965) showed both theoretically and through empirical data, that to optimally "pack together" a number of service centres and their service areas, the hexagon is the geometric shape that allows the greatest amount of packing into an area. At the same time the hexagon minimises movement to or from the boundary, and boundary length (Haggett 1965:49, cited in Hodder and Orton 1976:56).

Lösch (1941) departed from Christaller's central place hierarchical model. He elaborated a more flexible model, finding that first of all there is a tendency towards a concentration of sites into sectors separated by less dense sectors. Second, that there is a tendency towards an increase in size of sites, proportional to distance from the central site. And third, there is a tendency towards the location of smaller settlements halfway between larger ones (Clarke 1977:24).

Lösch's model as stated appears to be much more flexible and it involves a system in which neither size or social function of the central place needs to be directly or hierarchically related. In this model size, social role, prestige, power and even economic roles of central places are independent but not exclusive from each other, coinciding on occasions. These characteristics, however, pose some conceptual problems for the rank-size rule of Maya centres. In this sense the main point of departure from Christaller's model resides in his notion that centres which are functionally equivalent, need not be of the same size, nor must larger central places replicate all of the functions of smaller ones (Ball and Taschek 1991: 157).

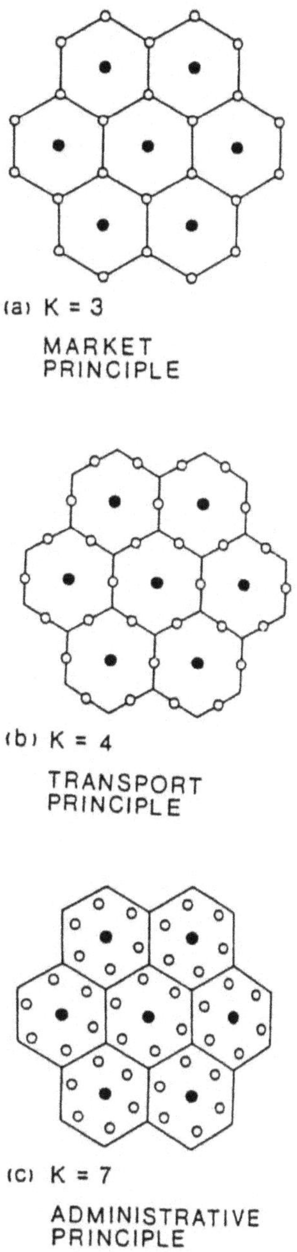

Figure 1-1 Central Place patterns of hierarchical distribution (From Inomata and Aoyama 1996)

Nearest Neighbour Analysis

Nearest Neighbour Analysis differs from Central Place Theory in the sense that it is aimed towards the consideration of the relationships of sites to each other rather than defining their putative regions of control (Hammond 1974:322). These types of studies fall within the realm of Point Pattern Analysis. The randomness of spatial patterns constitutes the underlying assumption of this type of analysis. This randomness can then be used as a norm against which a particular spatial pattern may be contrasted. The three types of distribution more commonly used as "yardsticks" are (1) regular or uniform distribution; (2) random distribution, and: (3) clustered distribution. Figure 1-2 shows the gradation between these distributions. With regard to site distribution, the identification of a non-random regularity of spacing, can readily be explained in terms of competition between sites (Hodder and Orton 1976:30-31).

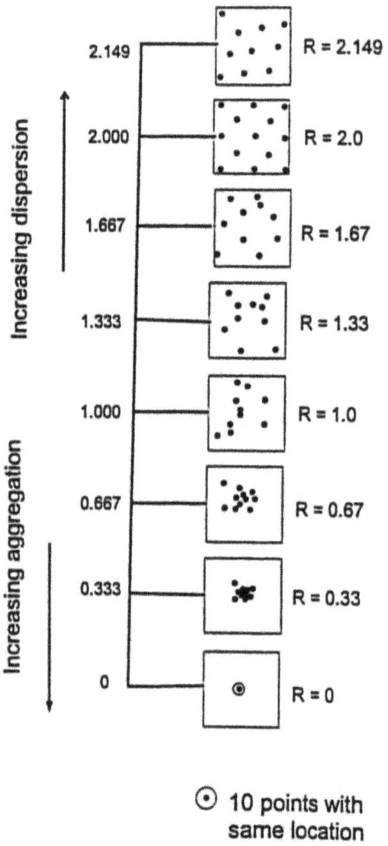

Figure 1-2 Nearest neighbour distribution types (From Adams and Jones 1981)

The estimation of nearest neighbour distance (**R**) is obtained through the following formula:

$$R = rA / rE$$

Where:

rA = average distance between a point and the point closest to it

rE = expected average random distance

In turn the expected average distance is obtained through the following formula:

$$rE = 1/2\,(\sqrt{N/A})$$

Where:

N = number of points

A = area of the map

In these analyses an **R** = 1.0, will indicate a random pattern, while the highest possible dispersion value is **R** = 2.15, and fractional values considerably below 1.0 will indicate an aggregated pattern. A dispersed pattern can reflect two possible phenomena: a spatially uniform resource base, or the internal competition between functionally similar central places. On the other hand an aggregated pattern can reflect either a strongly localised resource base, or extra-regional influences that determine location of centres based on strategic criteria (Adams and Jones 1981:314-315).

The Gravity Model

Originally derived from Newton's gravity law, the gravity model has been applied in archaeology to gauge the level of interaction between centres. Hodder and Orton however, caution that although higher levels of interaction are more likely between nearest centres than with any other similar centres located further away, attention should be set on size and importance of the sites. Larger centres tend to attract interaction from larger areas (Hodder and Orton 1976:188), taking this into account one should expect the interaction to increase proportionally to the size of the centre, and to decrease with distance. The amount of interaction between two places (marriage, alliances, royal visits, warfare, etc. [I_{ij}]) is obtained by the equation (Waters 1995:179):

$$I_{ij} = k\,((P_i)(P_j) / (d_{ij})^2)$$

Where:

k = is a constant depending on the phenomena

P_i and P_j = the population of i and j

d_{ij}^2= the square distance between i and j

A derivation of this equation aimed towards the calculation of the breaking point (boundary) between the service areas of two adjacent centres of different size, was developed by Reilly (1931, cited in Hodder and Orton 1976:188), where D_{xj} is the distance from breaking point (x) to centre: $D_{xj} = d_{ij} / 1 + \sqrt{P_i/P_j}$

With this equation the area of maximum influence of a centre can be predicted according to size and location of a neighbouring major centre, thus in this way complementing the nearest neighbour analysis.

All of the above described models have been and are still being applied in lowland Maya archaeology with different levels of success and will be discussed in the following chapter.

TWO SETTLEMENT PATTERNS IN THE MAYA LOWLANDS

Introduction

In his presentation of Spatial Archaeology Clarke deplored the state in which settlement studies were at that time. He argued that the vast majority still tended to be static and disjoint typologies without any attempt to give an account of the working principles that gave form to those patterns (Clarke 1977:7). In Lowland Maya archaeology the study of settlement patterns did not become a research problem until the explorations at Uaxactun by O.G. Ricketson Jr., in 1926-1931, when he extended his survey 1 mile beyond the centre in a cruciform pattern (Ricketson 1937). Prior to this it was the main centres, which received the full efforts of archaeological research, but without any conscientious attempt to link them together. It is not my intention to give an extensive account of the development of settlement pattern studies in the Maya Lowlands, excellent reviews can be found in Hammond 1972, Ashmore and Willey 1981, and Willey 1981. My dealing with this topic will be limited to the presentation of what I think have been the substantial contributions to the study of settlement patterns as a background to my own approach.

The most influential figure for the development of settlement pattern studies in North American archaeology was without any doubt Julian H. Steward. It was he who first suggested that in order to obtain insight into the behavioural patterns of societies a systematic approach to settlement patterns was necessary (Steward 1937). Steward maintained that through the study of settlement patterns research would advances in the study of cultural adaptations to the environment (Willey and Sabloff 1993:176-177; Ashmore 1981:10). In 1954 Gordon R. Willey, whom after his successful investigations of the Viru Valley in Peru (Willey 1953), introduced the systematic study of settlement patterns to the Maya Lowlands. Willey's settlement study focused on issues regarding site location, population size, sociopolitical organisation, and urban processes, throughout a time sequence starting at the Middle Preclassic to the Terminal Late Classic. As a result, settlement pattern studies were directed towards two broad classes of problems: 1) the ecological issues, which involved the interaction of individuals with their natural environment, and 2) the social-political issues involving the interaction between individuals (Ashmore 1981:4-11, 38).

William R. Bullard in 1958 carried out an extensive survey of more than 250 square kilometres throughout the northeastern Peten. His specific objectives were aimed towards the definition of the types of ruins that occurred in this region and how they were distributed over the countryside. Bullard also wanted to define the types of groupings in which these ruins occurred and what their relationship was to the larger ceremonial centres. Bullard's survey yielded a three-tier settlement structure: Clusters, Zones, and Districts, with the Districts being aggregates of Zones, which in turn were aggregates of Clusters (Bullard 1960:355, 367). Sanders (1962, 1963) proposed a similar scheme for the Chontalpa region in Tabasco.

Towards the latter part of the 60's and beginning of the 70's settlement pattern studies began to shift from site typologies to site function, deriving in issues involving the settlement implications for social organisation and population density (Haviland 1968, 1969, 1970; Kurjack 1974). Ashmore (1981) observed that these issues were directly related to the working assumptions that settlement patterns can yield information on human behaviour, and the number and/or density of the population occupying the landscape. The first assumption has already been initially dealt with in Chapter One, in which I argued that the level of analysis (individual structures, community layout, and inter-community patterning) determined the methodological approach. Hence the crucial aspect in reconstructing human behaviour lies in the specific functional interpretation that is given to the archaeological feature.

A problem that most of these interpretations encounter is the lack of a strict one-to-one correspondence between forms and functions. Nevertheless, Ashmore noted the importance that these functional interpretations have for the reconstruction of human behaviour, and argued in favour of establishing formal typologies for them. Spatial unit typologies, however, must include both an inventory of the elements that compose the settlement pattern, and the framework on which to articulate these different elements (Ashmore 1981:39-41, 43).

The reconstruction of population figures is largely determined by these formal typologies (e.g. houses, sleeping spaces, kitchens, etc.). Directly related to this is the generalised assumption that there is a direct correlation between population size and the observable quantities of archaeological structures. In general terms the house is considered as the counting unit, which in turn is transformed to a numerical ratio (e.g. 5.3 individuals per house) (Ashmore 1981:40). The main flaw in this reconstruction, however, lies in the uncertainty of proving the contemporary occupation of house mounds from surface observations.

Despite this difficulty, population figures and densities for various Maya centres have been estimated. In the process it has become clear that it is not possible to establish a uniform occupational pattern for all of the Maya lowlands due to the regional variations, which in part are construed by the characteristics of the landscape. These regional variations will be dealt with in the following section. All in all, nowadays it is generally accepted that the total population during the Late Classic Period in the Maya lowlands reached a seven-digit figure (Willey 1987:113).

Tikal, for example, represents the biggest centre of the southern lowlands. The surveys conducted at this site have established that its core area extends over 4 sq. km. A 6-km radius was traced from its core based on the presence of archaeological mounds, comprising a total area of 120 square kilometres. Haviland (1985) estimated a population of around 72,000 during the Late Classic. This means that the population density for Tikal was in the order of 600 persons per square kilometre. For Yaxha and Sak Nab, which are two other Peten centres, Puleston and Rice offer an estimate of 200 and 500 persons per square kilometres respectively. In the northern lowlands R.E.W. Adams (1981) estimated a population density of 500 persons per square kilometre, and for the area surrounding Becan, Rio Bec, and Xpuhil, Kurjack and Garza (1981) estimate for the area surrounding Dzibilchaltun a population density of up to 2000 persons per square kilometre.

It becomes apparent then, that all of these population figures were estimated taking as point of departure the main centres, from which it follows that the centres are seen as the nuclei of population aggregation. Consequently the study of settlement patterns has focused on explaining the spatial structuring of the various categories of sites from the core to the periphery. Puleston (1974) however, advanced the idea of surveying the areas between the major centres. In this context the explicit definition of centre was deemed necessary in order to elaborate the theoretical framework on which to mount the explanatory models that would account for the spatial distribution of sites over the landscape.

Nowadays the debate between ceremonial versus urban function for Maya centres has died down. The consensus points towards considering Maya centres more like true cities, because not only were they the foci of a variety of political, ritual, commercial and intellectual activity, but they were also actual residential loci (Ashmore 1981:16, 55). Willey has defined the centres as the aggregates or nucleated arrangements of "special purpose structures" such as pyramids, great platforms, plazas, palaces, and other administrative buildings that were the axis of political-religious life of the Maya. These centres were by no means devoid of residential units, but rather were surrounded and intersected by them, for which they corresponded more or less to a city (Willey 1987:113).

Functionally speaking then, the Maya centres were the framework on which the various social interrelationship networks were woven. These networks were based not only on kinship ties, but also on networks that surpassed kinship relations, such as economic and religious systems. They attracted individuals from all walks of life to actively participate in the various ritual ceremonies as well as in the exchange of goods and services (Morley, Brainerd, and Sharer 1984:210).

There is, however, a great deal of variation with respect to actual size of the various centres. These can range from those relatively small centres like Yaxha and Sak Nab, to the gigantic centres of Tikal and Calakmul, thus giving rise to a continuing debate on the criteria used to establish a hierarchical scheme between major centres. The hierarchical structure between sites can be reflected in its geographic distribution. It is expected that the main centre will be surrounded by other minor population loci, which in turn will be surrounded by residential units scattered around them, all of these usually within a 5 to 6 km radius from the major centre. Accordingly, the distance that lies between one major centre and the other should be of around 30 to 20 km. Throughout this extension of land, a series of minor centres are located thus giving form to the settlement network (Willey 1987: 128-130).

The analysis of settlement patterns in the Maya Lowlands is structured around the concept of site hierarchy. The criteria to consider a centre of first, second or third order vary according to the emphasis set on different material-culture aspects, or combinations of these. In general terms, based on the nature of the data used these criteria may be divided in two approaches:

- That, which has focused on spatial models to gauge the extension of the political-economical sphere of influence of the primary centres in relation to their distribution over the landscape (Hammond 1974, 1991; Adams and Jones 1981; Adams and Smith, 1981; Sanders 1981; Freidel 1981; Marcus 1983; Ball and Taschek 1991; Inomata and Aoyama 1996).
- The epigraphic approach, based mainly on the interpretations given to the presence/absence of emblem glyphs contained in the monuments of the various centres (Barthel 1968; Marcus 1976; Adams 1986; Mathews 1991, 1995; Martin and Grube 1995).

In the former models the criteria to establish site hierarchy varies considerably. The rank-size rule of centres has been essentially determined following two different parameters: the quantification of architectural elements and/or volumetric analysis (Turner, Turner, and Adams 1981; Adams and Jones 1981; Inomata and Aoyama 1996), and the complexity of road networks (Kurjack and Garza 1981). I will briefly review these approaches.

Spatial Models

Central Place Theory is perhaps the most widely used approach in the study of settlement patterns. In most cases however, Central Place Theory treats centres as having equal political status, despite acknowledging the existence of a rank-order between them (Hammond 1974:318).

Closely related to Central Places is a commonly used method to determine the influence area of centres through the construction of Thiessen polygons (e.g. Hodder and Hassal 1971; Hammond 1987; Mathews 1988, 1991). Thiessen polygons are created by drawing perpendicular lines half way between major centres, the polygons thus created can be seen as defining the areas which could most efficiently be served by the centres. This method is useful to the extent that it provides an initial territorial delimitation. However as mentioned above, its most critical limitation lies in the fact that these polygons give equal weight to centres of different size. Hence larger centres will not appear with a larger service area (Hodder and Orton 1976:59-60).

Ball and Taschek (1991), addressed this problem by applying Lösch's more flexible Central Place model to their Upper Belize Valley data within the context of the segmentary-state societies originally proposed by Fox (1977). In brief the segmentary-state can be described as being essentially decentralised, where the monopoly on wealth and power by a central figure is limited, and dispersed across subsidiary centres, which duplicate at a reduced scale the functions of the centre (Ball and Taschek 1991:159-160). In accordance with Fox's typology, centres were considered in a rank-order scheme as either regal-ritual centres or regal-ritual cities. The basic difference between these two is established by the nature of the demographics. The latter is understood in terms of the concentration of a large, dense, and socio-economically diverse resident population, which constitutes the hallmark of the city and represents the fundamental difference between both types (Ball and Taschek 1991:156-157)

Perhaps the most cited limitation to the applicability of Central Place Theory to archaeology lies in the very nature of its conception. Central Place Theory was originally proposed to address economic issues within a market-economy. Smith (1974, cited in Inomata and Aoyama 1996:292) observed that this theory could only predict the distribution of retail-market centres and not wholesale markets or administrative centres. Thus by nature, hindering its applicability in predicting or explaining site distribution without an adequate knowledge of market function. Inomata and Aoyama discard this limitation arguing that the essence of Central Place Theory (according to Christaller's model) can be utilised in the context of the spatial patterning that reflects the economy of cost of travel and transport. Thus addressing central places as "locational decisions for nodes of human activities [which] are made in general to minimise energy expended in movement" (Inomata and Aoyama 1996:292).

The application of Central Place Theory in archaeology, according to these authors, should be aimed towards the identification and assessment of the factors involved in the distribution of centres rather than on the prediction of the location of unknown centres or the description of the observed settlement patterns. Inomata and Aoyama emphasise that in order to be a more effective explanatory model Central Place Theory should be limited to a regional-level analysis focusing on the political and economic interactions between the centres and the local populations, instead of an inter-regional analysis.

Inomata and Aoyama analyse under the light of Christaller's Central Place Theory model the archaeological settlement data dating from the Late Classic period, from the La Entrada region located in northwestern Honduras. They establish the hierarchy of central places on the basis of the size of the centres using as a yardstick the architectural volume of the centres' main group. Christaller's model, however, is strongly based on the theoretical considerations regarding the circulation of people and goods, for which according to the authors relying solely on the settlement data available does not warrant the thoroughness of the results. Therefore, they support this data with the distribution of obsidian as an exchange good (Inomata and Aoyama 1996:293). The settlement and obsidian distribution data from the La Entrada region suggested to them that Christaller's market and administrative principles (Figure1-2) were the most important determinants of the spatial patterning and show a strong tendency to minimise the cost of travel and transport.

However, a problem that becomes apparent in Inomata and Aoyama's approach is the scale at which they are working. From the sample of 37 sites excavated only the centres of El Abra and Los Higos fall in the highest tier of their ranking system. These centres, however, would only rank as tertiary or secondary sites in an overall centre hierarchy. From these authors accounts it seems that the La Entrada region would have been part of the greater Copan region (Inomata and Aoyama 1996:300-301). Let us recall that these authors stress that the applicability of Central Place Theory should be confined to the regional level. Therefore the distribution of central places in the La Entrada region should be addressed considering the greater Copan regional settlement system as a whole, and from there identify the sub-regions ("zones" in Bullard's classification) and how they are linked to it.

Hammond (1974) pointed out that at first glance a non-uniform distribution of major centres was apparent in the Maya Lowlands. The distances between major centres ranged from 30 km around the Maya Mountains, to approximately 20 km in southern Peten and southern Campeche, to 10 km or less in northeastern Peten, where a clear *core area* was apparent. He proposed however, that this lack of uniformity was non-random and that it must reflect the relationships between the ceremonial centres, their rulers and their subject populations (Hammond 1974:320-321). Hammond makes use of Nearest Neighbour Analysis to better assess these data. As already mentioned in the previous chapter, Nearest

Neighbour Analysis is better suited to deal with the relationships between centres. Hammond's results yielded a regular pattern of centre distribution, with three apparent sub-areas:

1. The south Peten-upper Pasion River area comprising 25 centres with a mean nearest neighbour distance of 18.4 km.
2. A north Peten and south Campeche area comprising16 centres with a mean nearest neighbour distance of 15.1 km.
3. A northeastern Peten and adjacent Belize area comprising 15 sites with a mean nearest neighbour distance of 10.4 km.

These results clearly showed that there is a trend towards shorter nearest neighbour distances in the core area (Hammond 1974:322-326).

Again as in the case of Central Place Theory, one of the major limitations of Hammond's Nearest Neighbour Analysis is the absence of differential weighting between centres. Adams and Jones (1981), aware of this flaw, propose that along with the spatial patterning of the centres, a rank-size rule based on courtyard counting and volumetric assessment should be used in order to objectively assign hierarchies. Volumetric assessment was deemed by Turner, Turner and Adams (1981) as the only objective quantitative means of establishing site hierarchy to all centres, which would be independent of style, carved monuments and excavated data. The method consists in combined measurement of the quantitative and qualitative dimensions of the architectural components of a centre. It is in essence a three dimensional geometric model. In this model the two-dimensional facet is represented by the quantitative aspects of the architectural elements (open space areas, total mass, access or circulation elements, number of functionally similar facilities). The depth or third dimension describes the qualitative and/or cultural aspects (space relationships, materials used, ornamentation, aesthetic embellishment, etc.).

These facets are reduced to a numerical equivalent that can be weighted and added together. The sum of these, results in a three dimensional volume value, and in turn the cube root of this value will give the numerical value of the rating of a specific centre (Turner, Turner, and Adams 1981:72-75).

A four-tired scheme was obtained which according to the authors, should reflect the political, economic, and demographic hierarchies between centres (Adams and Jones 1981:302-307). Based on architectural and ceramic styles, and spatial contiguity, as well as on sample size and reliability of the data Adams and Jones propose four regions. The Tikal region of northeastern Peten; the Calakmul region located on the boundary of north Peten and Campeche; the Rio Bec region of Campeche and Quintana Roo; and the Chenes region in Campeche and Yucatan (Adams and Jones 1981:308). By plotting the centres on a double-logarithmic scale graph, three types of site distribution which reflect the state of regional development of the political geography can result:

1. A downward-sloping straight line, which means that there is a continuous progression of centre size.
2. A concave curve, which marks a primate distribution suggesting the existence of a single major centre and a number of smaller sites.
3. A convex curve, signalling a plural distribution that suggests a distribution with several large places of equal size and a paucity of smaller places.

The first type of distribution is typical of large regions or of those that have had long periods to evolve. On the other hand the second or primate distribution represents the opposite of the first, that is a system in which few simple forces have operated, and is typical of smaller or younger regions. The third or plural distribution is typical of a rank-size distribution with even longer periods of evolution. This type of distribution is best characterised by the concept of "closure" which is the extent to which social transactions take place within a system rather than between systems (Adams and Jones 1981:310-311).

Adams and Jones' results suggested a developmental sequence for the four regions. The Tikal and Calakmul regions showed a more or less standard log-normal distribution characteristic a rank-size rule exhibiting a size continuum, while the Chenes and Rio Bec regions show a strong plural distribution. Adams and Jones find that these patterns indicate that the former regions were economically mature and Tikal and Calakmul were the undisputed dominant centres with a balance between external and internal growth. On the other hand the Chenes and Rio Bec regions also had an early development, which was truncated with the collapse of the southern centres only to develop further afterwards (Adams and Jones 1981:301, 319-320). In all cases the authors view a tendency towards dispersion which is akin to the feudal model of Maya society proposed by Adams and Smith (1981), and which will be discussed in more detail.

The integration and articulation of the above spatial models to the settlement system has been approached by means of the proposal based more or less on a feudal model of social organisation (Adams and Smith 1981; Sanders 1981; Freidel 1981). An interesting aspect about these models is their strong reliance on the ethnographic analogy outside the Maya area, from such remote locations such as Europe, Japan, and East Africa.

Adams and Smith's feudal model is characterised by the diffusion of political power from the centre down with a well defined hierarchical structure within the elite, that in turn will find its counterpart in the site hierarchy. In this latter hierarchical structure a tendency to duplicate the functions of the primary centres by the minor centres at a lower level can be observed in the architecture according to the authors (Adams and Smith 1981:343-346).

Sanders (1981) proposed that the structure of settlement patterns was determined by the complex interrelationship that is given between agricultural technology, the natural environment, and the sociopolitical organisation. In this model the increase and dispersion of the farming populations since the Preclassic resulted in an increase of the demographic density, which along with the development of more intensive agricultural techniques allowed the emergence of densely populated centres such as Tikal. Sanders, although emphasising the relevance of the ecological factors in such developments, also recognises that the choice of a population is based mainly on motivations of political control. In this manner Maya society is best described as having an eminently hierarchical structure, in which individuals occupied well-defined tiers on the social pyramid. Akin to this hierarchical structure was the settlement hierarchies whose main objective was the administration and management of the agricultural resources (Sanders 1981:354-359).

Freidel's (1981) pilgrimage fair model views the process leading to the emergence of centres as a social invention. This model is based on two operative principles: (1) the controlled flow of the individuals to the centres through religious motivations (pilgrimage), and (2) as a mode of integrating groups of individuals by providing a medium to exchange goods and services (fairs). These fairs involved periodic markets that would articulate the local economies into a wider network. As a result a distinctive form of government would emerge, which according to Freidel, gave an appearance of equality in a non-egalitarian society. In this model the secondary nobility or feudal lords that were settled in the periphery would administer more efficiently the marginal domains of the polity if they were viewed by their subjects as belonging to the local communities, rather than as being imposed by the central power. This model differs from the two previous ones in that it gives a greater importance to hand-craft production and its redistribution, although it is stressed that the former was probably confined to the household level (Freidel 1981:371-382).

The archaeological data provides certain support to these models, there is a clear gradation in terms of size and complexity in the architectural remains stemming from the primary centres to the secondary and tertiary centres. This alone, to a certain extent, can be seen as a reflection of the parallelisms that existed between Maya society and the other feudal societies of the world. Both Adams and Smith, and Sanders see in the Maya centres a political function, in which the primary centres would house the royal residences. These in turn would be surrounded by their courtiers, and would constitute the core of political power from where central control would be wielded with the support of the secondary nobility residing in the subsidiary sites scattered across the periphery of the polity (Willey 1987:129-30).

The only example I know of the use of the Gravity Model in the Maya lowlands is that of Dunham (1990). Dunham applies Reilly's derivation of the original Gravity Model equation (see Chapter 1) to the problem of marginal development of centres along the boundaries of pre-existing primary centres. Among his principal objectives, Dunham's main thrust is centred on the identification of the implications that the process of marginal development had for the Classic Maya collapse within the context of segmentary states. Applying Reilley's equation Dunham reconstructs the boundary between the polities of Nim Li Punit and Lubantuun in southern Belize in order to address the development of Xnaheb, a small Late Classic site located between these two polities. His results suggested that the development of centres along the boundaries of pre-existing centres is a late event that takes place after secondary centres have first developed on the periphery. This process illustrates the weak political integration that the segmentary system has which eventually leads to its own demise (Dunham 1990:148-152, 641-643).

The Epigraphic Approach

This is a relatively more recent approach, which throughout the development of settlement studies has acquired more reliability. Its basic assumption is that only centres with inscribed monuments are considered for rank ordering, and within these the absence or presence of emblem glyphs and their frequency constitute the criteria to assign rank to a centre. Although Berlin (1958) is the first one to identify the emblem glyphs as a sort of toponym, Barthel (1968) pioneered the idea to use this glyphic element as a criterion of site hierarchy. Based on the occurrence of these glyphs Barthel proposed two kinds of relationships: a simple and a complex one. In the former the emblem glyph of a site **A** would occur in the inscriptions of a site **B**. The latter would involve more than two sites, in which at the very least the emblem glyphs of sites **A** and **B** occur in the inscriptions of site **C** (Barthel 1968:185). Furthermore, on the basis of his interpretation of Stela A of Copan, Barthel proposed the existence of four regional capitals. Each of these capitals was located in one of the four directions of the Maya cosmos (Figure 2-1): Copan, Tikal, Palenque, and a site at that time unidentified, which was later identified as Calakmul (Barthel 1968: 185-193).

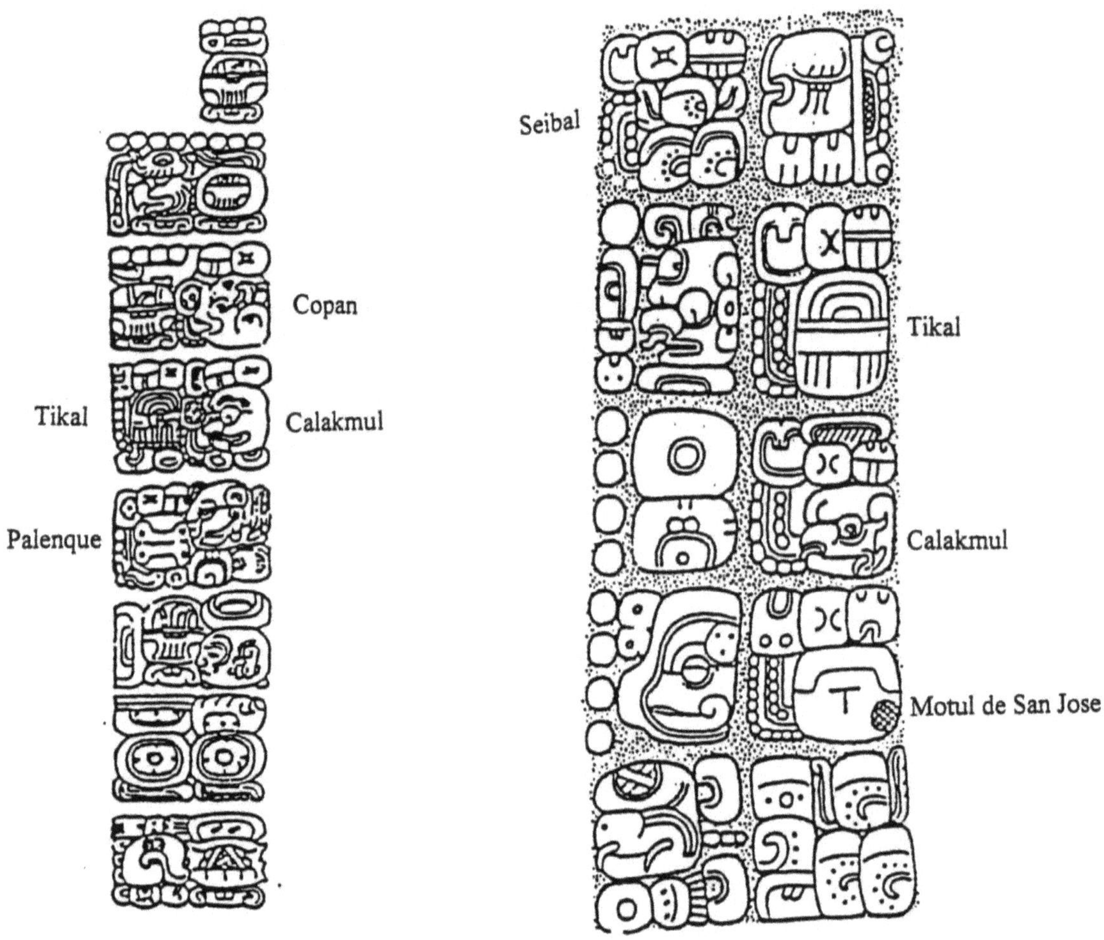

Figure 2-1 Regional capitals of the Maya World (a) Copan Stela A; (b) Seibal Stela 10 (From Grube and Martin 1998)

In a similar manner Marcus (1973,1976) proposed a hierarchical system in which the minor centres would include in their inscriptions the emblem glyph of the more important centres. Thus a four-tier hierarchy resulted from these criteria (in Morley, Brainerd and Sharer 1983:213):

1. Primary or first order centres, which have their own emblem glyph although the emblem glyphs from other primary centres may occur in their inscriptions. Examples of these are Tikal, Yaxchilan, Copan, and Palenque.
2. Secondary or second order centres are those which have their own emblem glyph and which will include the emblem glyphs of other primary sites in their inscriptions but not the converse. Examples of these are for the Tikal region Aguateca, Machaquila, and Naranjo; for Yaxchilan, Piedras Negras; Quirigua for the Copan region, and Pomona, for Palenque.
3. Tertiary or third order centres, were those that did not possess their own emblem glyph but have inscribed monuments in which the emblem glyph of a primary or secondary centre occur. Examples of these according to Marcus are Ixlu, Ximbal, and Uaxactun for the Tikal region, Bonampak and El Cayo for Yaxchilan, Pusilha for Copan, and El Retiro, Jonuta, Miraflores and El Tortuguero for Palenque.
4. On the lowest tier are included all those sites which do not exhibit any emblem glyph at all in their inscriptions (Marcus 1973:10).

Note that Marcus omits from the list the important centre of Calakmul and considers Yaxchilan as hierarchically superior to Piedras Negras. However, she would later modify this original scheme (Marcus 1976). As Barthel before her, Marcus sees Stela A of Copan for the Late Classic and Stela 10 of Seibal for the Terminal Classic as evidence of the existence of four regional capitals whose greater hierarchy argues in favour of the presence of political hegemonies in the Maya Lowlands (Marcus 1976:). Following this line R.E.W. Adams (1986) also proposes the existence of eight regional capitals in the Maya Lowlands during the Late Classic.

The phonetic decipherment of the emblem glyph (Lounsbury 1973; Mathews and Justeson 1984; Mathews 1985; Mathews1991) showed that this, as part of the titles included in the name phrase of rulers and royal nobles, denotes an equal political rank between individuals (Figure 2-2).

Figure 2-2 Components of the emblem glyph (From Mathews 1991)

In this context Mathews (1991) found unlikely the existence of regional political capitals. The advances in the epigraphic knowledge of the time, which revealed the nature of the relationships between centres, did not support then the idea of the existence of regional capitals. Mathews argued that if the emblem glyphs were referring to centres and dynasties of equal political status, then there were as many independent polities in the Maya Lowlands as there were emblem glyphs. The result was a total of around 60 or 70 autonomous polities during the Late Classic (Figure2-3) (Mathews 1991:29).

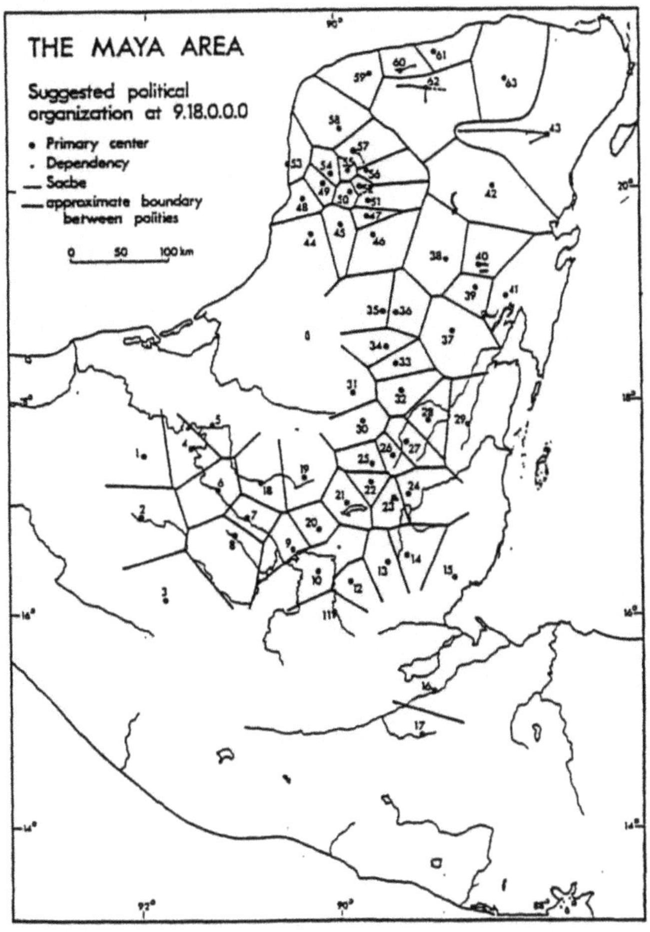

Figure 2-3 Identification of autonomous polities (from Mathews 1988)

The constant mention of warfare in the inscriptions between centres throughout the Classic Period is at odds with the idea of any sort of regional integration. As Grube and Martin noted, no unitary regional states could have tolerated such a frequent internal warring and still exist as such (Grube and Martin 1998:II-12).

Mathews' 'fragmented' political map of many autonomous polities is akin to the 'peer polity' model proposed by Renfrew (Renfrew 1982; Renfrew and Cherry 1986). In this scheme, although a centre may be able to gain prominence over other neighbouring autonomous centres, this prominence is short-lived and no real empire or regional state ever occurs. Houston has defined this form of political organisation as the 'weak state' in opposition to the 'large strong state' idea that prevailed with the proposed existence of the regional capitals. In short, the Maya 'weak states' are characterised by a friable structure with a limited control over its people and territory. Political power is centred around the figure of 'charismatic' kings, a bureaucratic administrative structure is weakly developed, and the power of the king will be dependent on his ability to attract and maintain the support of the provincial barons (Houston 1992, cited in Grube and Martin 1998:II-14-15).

However, more recent interpretations by Nikolai Grube and Simon Martin point again towards the existence of hegemonic political powers (Simon and Grube 1995; Grube and Martin 1998). Their findings, revealed data that was in direct conflict with the 'peer polity' 'weak state' paradigm, but at the same time fell short of corroborating the existence of regional states. In their analysis of the Classic Maya Lowland inscriptions, Grube and Martin found that there was enough evidence to suggest some forms of subordination between centres. For example, just as the subordinate relationships between kings and their vassals (*sahal)* may be inferred by the possessive form *u-sahal* "the sahal of" (Figure 2-4), the political subordination of one king to another can be attested by a similar possessive form *y-ahaw*, "the lord of" (Martin and Grube 1995:42).

In the same line of reasoning the presence of a different term that introduces a new character (always a lord or ruler from a different polity) reinforces this notion. This term is *u-kahi* also read as *u-kabhi* (Figure 2-5), which occurs in various royal inaugurations and has been translated as "under the supervision of" or "under the aegis of" suggests the direct intervention of a dominant king in the accession of another. More implicit references of direct supervision are the terms *y-ichnal* and *ilah*. The former has been glossed as "together with" or "in the sight of", and has been identified in three references to foreign supervision of accession ceremonies, while the latter, translated as "was seen" appears on a single example (Grube and Martin 1998: II-16).

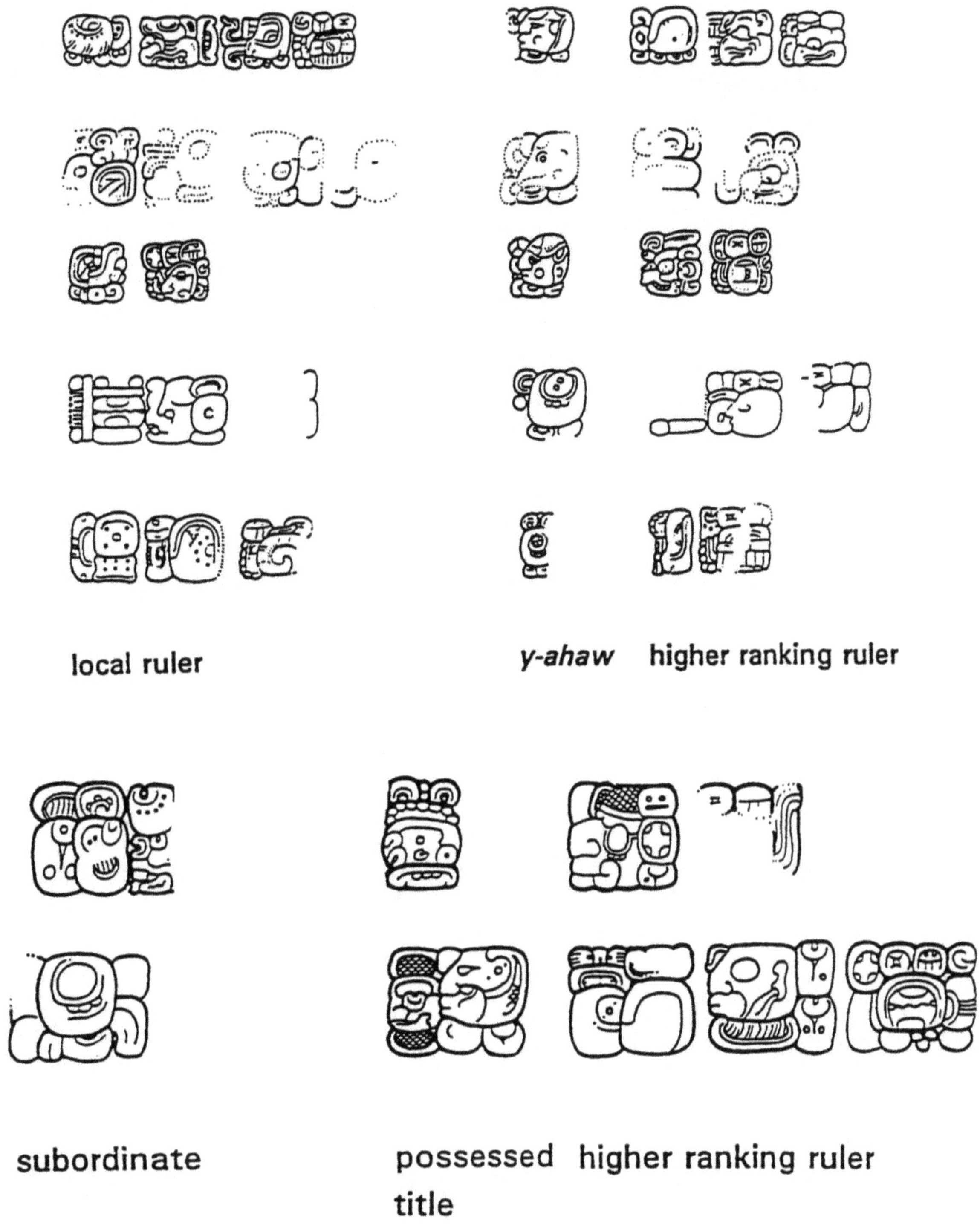

Figure 2-4 Titles that imply political subordination (From Grube and Martin 1998)

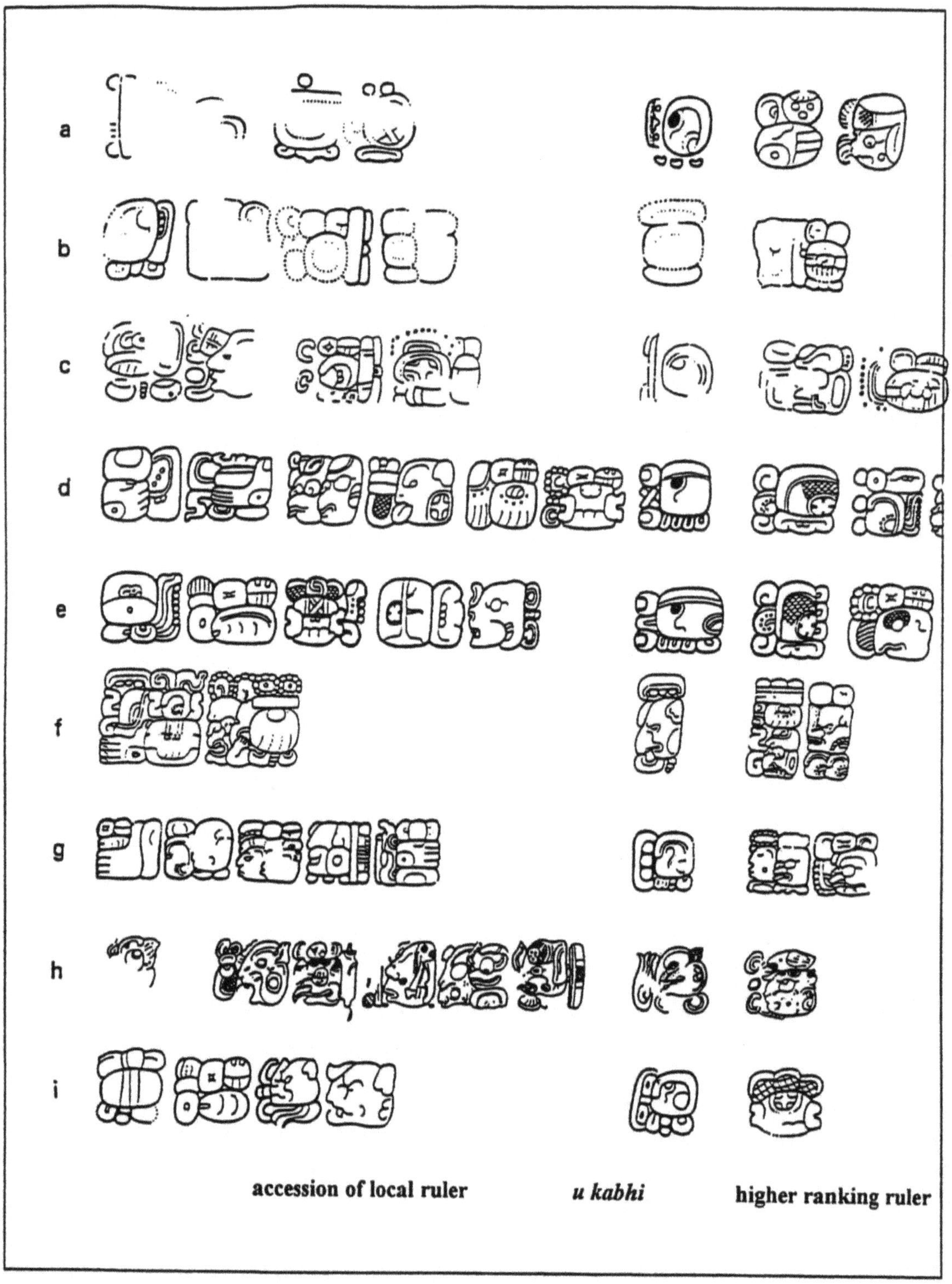

Figure 2-5 Expressions that imply political subordination (From Grube and Martin 1998)

From this analysis Martin and Grube were able to identify that some kingdoms were consistently more dominant than others, manipulating or at the very least intervening in the internal affairs of the smaller ones. The few explicit inter-state warfare records also seem to be consistent with the idea that war rarely took place between centres that usually had political ties; while at the same time political allies seem to share the same enemies. Many of these political subordinate relationships occurred between neighbouring kingdoms, but a substantial number of these crossed half the Maya Lowlands (Martin and Grube 1995: 44). Political influence and power was obviously not shared equally between states; rather Martin and Grube have found that there is enough evidence to propose the existence of two hegemonic superpowers: Tikal and Calakmul.

THREE LANDSCAPE ARCHAEOLOGY AND COMMUNICATION ROUTES

Introduction

Geography has once again supplied archaeology with the theoretical and methodological tools to assist it in the analysis of the spatial nature of the archaeological record. This time through the conceptual tenets of landscape theory that have provided the foundations of landscape archaeology. From these, Crumley and Marquardt (1990) have identified two types of structure that determine landscape: the **social-historical structures** and the **physical structures**. The former includes the political, legal and economic structures, as well as the administrative units necessary to implement these. The latter are those structures that are relatively out of direct human control (although imbued with societal meaning), such as climate, topography and geology (Crumley and Marquardt 1990:74).

Archaeology's concern with human behaviour goes hand in hand with the spatial analysis of the distribution over the natural and social landscape of the material culture and the people that created it. Hence, landscape archaeology is concerned with the study of the human imprint on the natural environment, where the cultural landscape reflects the interplay between technology, environment, social structure and the values of the society that gave it form (Green 1990:5; Trombold 1991:1).

It was within the context of landscape archaeology that Aliphat (1994) proposed that the location of the major known centres of the Upper Usumacinta region was strongly correlated to the natural landscape elements, giving priority to the strategic factors, in this case, channels of communications. The route systems used by a particular culture should then reflect something of its internal composition, value system (in terms of choices of connection), and mode of adaptation to the cultural and natural environment. Routes can either be formal routes: those that show evidence of planning and purposeful construction, or informal routes: those that have minimal or no labour directed to their creation or maintenance. The latter mostly consist of paths, trails and trade routes (Trombold 1991:3).

In this chapter I will deal first with the physical structures of the landscape and the effects that this landscape has on movement. I will leave the discussion of the sociohistorical structures for Chapter Five.

Description of the Physical Environment

The physical environment of the Upper Usumacinta region has been studied and thoroughly described by Aliphat (1994). Aliphat's study area extends from the origin of the Usumacinta located at the junction of the Pasion and Salinas Rivers, to Boca del Cerro. For the purposes of this study I have extended Aliphat's original area to the northwest beyond Boca del Cerro in order to include Pomona. The Lacanja Valley was also included; thus the total area covered in this thesis comprises some 9254 sq. km. I refer to the study area as the Upper Usumacinta Region (Figure 3-1), this definition is based more on the levels of site interaction present in the study area, than in its actual physical definition as in Aliphat's thesis. In this sense the study area represents what Taylor (1975) designates as a *functional region*, which is characterised by steady flows of spatial interaction and is defined by a distance decay effect (Taylor 1975:5).

The nature of the site interaction will be discussed in more detail in Chapter 6, for the time being it suffices to mention that the nature and levels of this interaction are well accounted for in the hieroglyphic corpus of the region. Figure 3-2 shows a map of the region with the location of the main sites. Although this site map includes both primary centres as well as secondary ones; it is by no means exhaustive since only those sites that have monumental architecture were considered. Some of these were visited and recorded by Maudslay, Charnay and Maler in the latter part of the 19th century and more sites are still to be discovered. The dense concentration of sites towards Boca del Cerro is probably more the effect of the recent surveys carried on by the Comisión Federal de Electricidad (CFE), Mexico's state owned electric power company. The CFE was conducting a series of environmental and cultural resources impact studies for a projected construction of a dam at Boca del Cerro. Archaeologist G. Delgado produced a map that comprised these sites and other minor sites that were not included in my study (Delgado nda). I will briefly summarise some of the main features of the physical environment analysed by Meave (1990) Aliphat (1994) and Giovannini et al (1998), which I consider of prime importance for the present study. I refer the interested reader to these works for a more detailed account. In this summary description the reader will notice that I am not including a discussion of the vegetation despite its obvious importance. This dismissal however is due to the lack of more extensive paleo-environmental studies in the region.

Topography:

The region under study is characterised by hilly terrain covered with dense tropical forest on the Guatemalan side, and with the exception of the Reserva Biosfera Ecológica Montes Azules, there are just a few remaining patches of tropical forest left on the Mexican side. The Usumacinta basin is bounded to the north and south by mountain ranges. The northern range located on the Guatemalan side of the river is known as the Sierra de Lacandon. This mountain range reaches an altitude of about 500 masl, with an approximate length, running from northwest to southwest of 75 km. With these characteristics the Sierra de Lacandon constitutes a formidable natural barrier towards the Peten. Located to

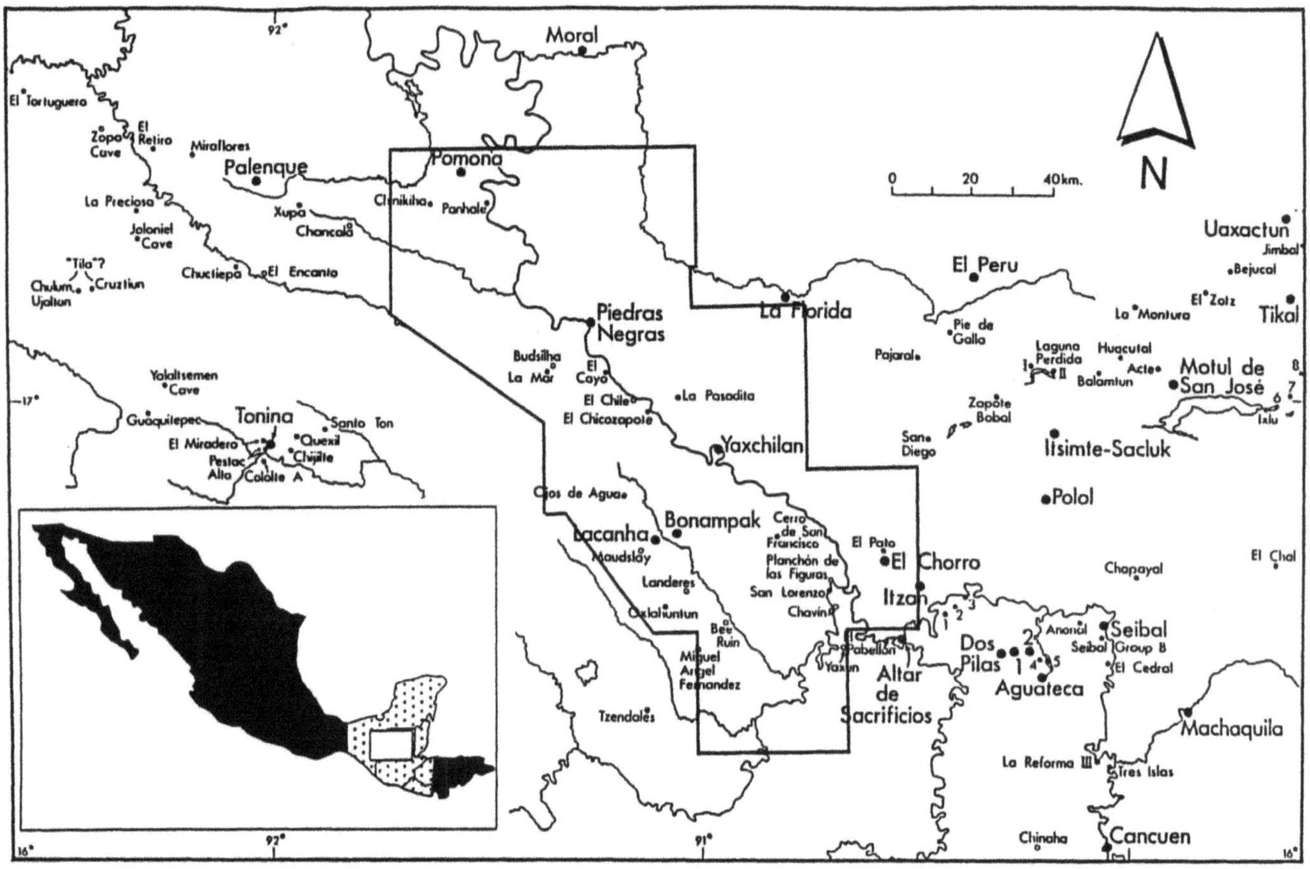

Figure 3-1 Map of the Western Maya Lowlands indicating the Upper Usumacinta Region (adapted from Mathews 1988)

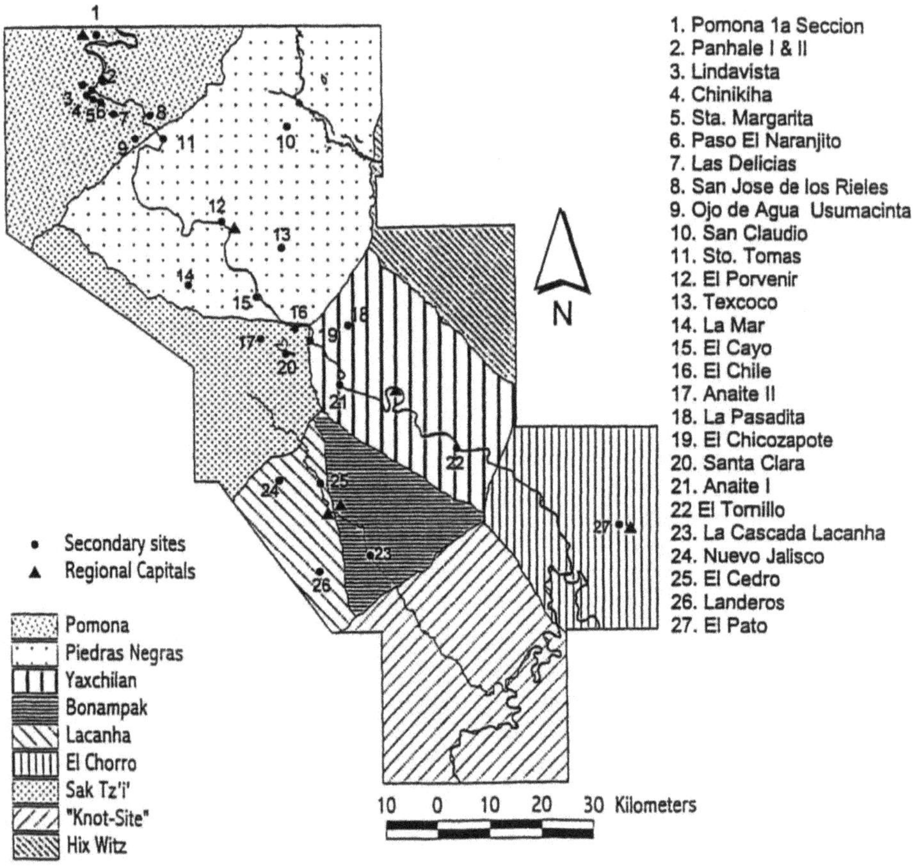

Figure 3-2 Archaeological sites included in the study

the south of the Usumacinta River is the Sierra de la Cojolita. It rises up to 700 masl, running parallel to the Usumacinta basin for approximately 70 km, and separating the Usumacinta basin from the Lacanja basin to the south. However, a series of natural passes formed by the faulting of the range structure connects both valleys. In between these two mountain ranges tracts of relatively flat lands are located running parallel to the Usumacinta. These lands are characterised by the presence of marshes, swamps streams, rivers, lakes and rolling hills (Aliphat 1994:21-22).

The Lacanja valley is bounded to the east by the Sierra de la Cojolita, and to the west by the Eastern Highlands. The floor level of the Lacanja Valley lies at an elevation of approximately 350 masl. It contains a series of karstic hills that can reach up to 50 m of elevation from the valley floor. The Lacanja River intersects this valley forming two plains slightly slanted towards the river. The western side is approximately 18 km wide, while the eastern side measures around 10 km in width. It is in this latter zone that the important Classic Maya centre of Bonampak is located, about 3.5 km from the Lacanja River (Meave 1990:22).

Climate:

Although exhaustive records of the weather fluctuations have not been kept due mainly to the lack of observation stations in general, high temperatures and high precipitation characterizes the climate of the Usumacinta basin. The mean annual temperature surpasses the 24° C, with an average high of 30° C (Aliphat 1990:44). For the Bonampak area, however, Meave reports slightly lower mean temperatures of 24.6° C, with an average high of 27.2° C during the month of May (Meave 1990:19). The annual mean precipitation is 2,200 mm in the Upper Usumacinta and 2,609 mm for the Lacaja Valley. An important factor to keep in mind in terms of river and land transportation is that precipitation is seasonal. The rainiest period occurs during the months of May through September with a monthly average of 400 mm. Aliphat notes that during the rainy season there is a drier period in the months of July and August, known as the *canícula*. On the other extreme February and March are the driest months with a rainfall of under 40 mm on average (Aliphat 1994:45). In the Bonampak area the annual mean precipitation reported by Meave is 2,609 mm, with the rainy season beginning in the month of May (136 mm) reaching its peak in August (468.8 mm), and concluding in January (111 mm). The driest month is March with an average precipitation of 34.5 mm (Meave 1990:19).

Drainage System:

Five major rivers and their tributaries form the hydrological network of the region: the Usumacinta, the San Pedro Martír, the Lacanha, the Lacantun and the Salinas Chixoy. Of these five only the San Pedro Martír is part of a separate network that connects the region with the Peten. All of these rivers are navigable for most of their course thus forming part of very important communication routes. Of the five rivers the Usumacinta is by far the most important for the region. It virtually constitutes the spinal cord of the whole region, and so more detail will be given to it in the discussion that follows.

Contrary to common belief the Upper Usumacinta River does not form a river valley proper since owing to its structural characteristic, it lacks a floodplain. This peculiarity is due to the fact that the course of this river does not follow a valley created by the natural folds of the earth's surface (syncline), but rather it runs along the faulted axis of an uprising fold (anticline) heavily eroding its core. Aliphat better describes this characteristic: "The river is in effect, trapped in the core of this structural feature and will cause deep erosion in a rectilinear pattern, rather than in a sideways pattern which characterizes other major sized rivers in the formation of sinuous or meandering river valleys" (Aliphat 1994:32). The Lacanja River follows a northwest to southeast course direction running for approximately 75 km until it reaches the Lacantun River that flows from the south. The Lacantun in turns follows a northeasterly direction up to the point where it joins the Salinas/Chixoy Rivers giving origin to the Usumacinta (Meave 1990:23).

The Usumacinta River crosses two very important physiographic regions within the study area. These regions control the nature of the river's course. Upstream from Yaxchilan the river leaves the Usumacinta anticline and consequently its course becomes more sluggish and less broken, as opposed to its course downstream. In this section, beginning from the Yaxchilan oxbow, the river is characterised by the frequent presence of gorges with steep escarpments. There are five major segments along the river, which run through steep canyons. Aside from the escarpments there are also a series of rapids, five in total. These rapids are due to the transcurrent faults that have cut the axis of the Usumacinta anticline (Aliphat 1994: 53, 56, 65).

During the rainy season the levels of the rivers and streams rise considerably. The fact that the Usumacinta does not have a flood plain makes these changes even more radical. September, October and November are the months when the Usumacinta rises the most; while during the month of May the river is at its lowest level river (Aliphat 1994:54-55). I will illustrate in the second part of this chapter, how these changes in the river's level can have an important effect in the travel possibilities.

Soils:

In pondering the strategic vs. the tactical factors for the location of human settlements, soils (in terms of quality and agricultural potential) always loom large as a determinant variable. However, the Usumacinta basin, lacking a flood plain proper, does not represent great agricultural potential. Despite this, two of the major centres of the region (Piedras Negras and Yaxchilan), and at least five important secondary centres (El Porvenir, El Cayo, El Chile, El Chicozapote, and El Anaite I), are located along its banks. Aliphat suggests that an examination of the soils found in the vicinity of these sites is deemed necessary in terms of their agricultural potential. His analysis shows that none of these sites is located near an exceptional soil type in terms of agricultural productivity. Aliphat identified eight distinct soil series in his study; of these the most abundant are the Lacandon Series that represents over a third of the whole area. These series are characterised by shallow but fertile and well-drained soils. This means that although not prime agricultural soils they are still very good soils with only minor limitations (Aliphat 1994:122, 130, 135). Meave briefly described the soils of the Lacanja Valley in the vicinity of Bonampak. His description characterises these soils as similar to those of the Lacandon Series (mainly rendzina, which are shallow but well drained with moderate fertility). However this identification is based more on deduction rather than on empirical observations (Meave 1990:24). Giovannini et al. (1998) have conducted a more thorough study of the soils around Bonampak and along the Lacanja Valley, and have found that the most important soils present in the region are Rendzina, Ferric Luvisols, Eutric Nitosols, and Lithosols. As in the case of the Upper Usumacinta the higher percentage of soils for the Lacanja Valley are represented by the Rendzina. However, these studies also revealed that around the sites of Bonampak, Lacanha, and La Casacada Lacanha there is a high representation of Luvisols, which are by far prime soils for agricultural production.

Aliphat concludes that the location of the archaeological sites in the Upper Usumacinta valley is highly correlated with the river itself rather than with the proximity to any significant availability of optimum agricultural soil. This seems to be also true for the inland communication routes as the location of the sites La Pasadita and Anaite II -- both of which are found on the edges of swampy and poorly drained areas (Aliphat 1994:136). On the other hand, the location of the two major centres in the Lacanha Valley (Bonampak and Lacanha) seems to be more correlated to the vicinity of prime agricultural land. Giovannini et al. study shows that Bonampak is not only located in a region that is more than suitable for maize agriculture, but also that the pockets of fertile deep soils located close to the site, provides the potential for large-scale cacao production. Such a possibility is supported by the mural paintings of Bonampak where bundles of five by eight thousand cacao (*5 pi kakaw*) are offered in tribute (Miller 1997; Govannini et al. 1998).

Travellers to the region nowadays find their journey facilitated for the most part by the modern means of transportation (asphalt or dirt roads, airstrips, outboard motor boats). Their visit to the various archaeological sites of the regions is for the most part uneventful, with the exception of the occasional bandit attacks. This ease of travel was not the case just a couple of decades ago, when the options where limited to mule and a few airstrips, and even earlier going into the region meant walking and canoeing for the most part. To further illustrate the limitations that the physical structures of the landscape poses to travel let us turn now to some earlier accounts when travel was still made mostly by foot and canoe.

Traversing the Region, the Historic Accounts

Although what Trombold has defined as formal communication routes have not been indisputably identified, communication routes throughout the study area are by no means haphazard. The knowledge of the natural inland passes and the location of river rapids and portages was and still is essential for the success of any incursion into the region. To this respect we have a wealth of information in the accounts of the sixteenth century Spanish conquistadors and the late nineteenth century-early twentieth century scientists-adventurers.

The travellers' accounts specify that they used the old *monteria* trails cleared by the men seeking precious woods, chicle, and rubber in the Lacandon forest. Nevertheless as we read their accounts it becomes obvious that due to the abruptness of the terrain and the numerous natural obstacles these trails followed close to, or better yet were the same routes of pre-Hispanic times. This assumption is reinforced by Charnay's and Maler's mention of the many archaeological ruins lying next to these trails.

Córtes' punitive excursion to Honduras led him to the outskirts of the region as he arduously passed through Tabasco and Campeche in 1525. It was then, when Córtes first heard mention of the Chiapas lowlands and a small but bellicose tribe that inhabited an island in a big lake located in a region known as *Lacam-Tun*, the Lacandones. Pedro de Alvarado, then Governor of Guatemala, learned about his former captain's difficult journey. Alvarado also knew about the existence of the Lacandones, so he organised an expedition with the double intention to join Córtes' forces and to conquer the Lacandones. Alvarado was unable to reach Cortés, although de Vos (1988), in his magnificent treatise on the conquest of the Lacandon forest refers to a couple of legal documents and a letter written by Alvarado himself attesting to the expedition. In his letter Alvarado mentions how despite all his efforts it was impossible for him to join Córtes' forces due to the impassable marshes and rivers that lay in between (de Vos 1988:46-48).

Five years later in 1530, the first military incursion into the Lacandon forest took place under the command of Alonso Dávila. Dávila was one of the captains of the *adelantado* Francisco de Montejo conqueror of Yucatan. An eyewitness related the details of this entry; Alonso de Luján to Gonzalo Fernández de Oviedo y Valdés, and are commented upon extensively by de Vos. Montejo during his campaign to conquer Yucatan wanted to establish his operational base in the province of Acalan in Campeche, which during that time had a reputation of being prosperous and densely populated. Montejo held a meeting with Juan Enríquez de Guzmán, then Governor of the province of Chiapas. Guzmán agreed to aid Montejo in his bid for Acalan and suggested taking a route through Chiapas, to which he would provide him with arms, provisions, and more importantly Indian guides. Due to illness, Dávila instead of Montejo led a small force of eighty men through the Lacandon Forest. This small army of conquistadors departed from Ciudad Real (San Cristobal de las Casas) accompanied by the Indian guides who led them for the first thirty "leagues", after which they returned with the excuse that they no longer knew the region or spoke the local language (de Vos 1988:48-50).

Dávila continued his advance to the jungle and after walking for a considerable (though unspecified) distance they reached the banks of a big lake with a small rocky island near the shore; they called this lake Laguna de Lacandon. On the island Dávila's forces found a village of about sixty well-constructed houses. According to this description and those from later entries, the lake, island and the village match the description of the *Lacam-Tun* capital that Cortés had heard mentioned on route to Honduras. Blom in 1950 identified this lake with the present day Laguna Miramar. Further more, Blom noted the remains of platforms and ancient buildings on one of the island close to the shore (de Vos 1988:49-50).

Dávila's forces attacked the inhabitants of the island in search for gold. They found none; but they did manage to capture some Indians who served as guides for them to cross the region on route to Acalan. It is interesting to note that these prisoners according to Luján's account claimed to know the route to Acalan. This statement is important since it suggests that contact between the Upper Usumacinta region and the coastal plain continued even after the Classic Maya collapse. It also suggests that the communication routes used during the time of European contact may have been the same as in prehispanic times. Dávila's march continued and after walking through approximately thirty "leagues" of marshes and abrupt terrain they reached a large fast-flowing river, which had to be the Usumacinta (Figure 3-3). They reached a small settlement on the banks (the left bank) of the river, where they were well received by "friendly" natives who provided them with canoes and offered to guide them downstream. A short time later they reached treacherous waters that virtually horrified them (the Anaite rapids?). I am inclined to think that the natives expected the Spaniards to perish at the rapids since according to Luján's account they survived them miraculously. After three more leagues of treacherous waters (the San Josecito rapids?), they reached a more peaceful stretch of the river finally arriving to the Chontal town of *Tanoche*, presently known as Tenosique. From this point Dávila and his army continued their journey heading east, crossed the San Pedro Martír River after which they reached their final destination, the province of Acalan (de Vos 1988:50-51).

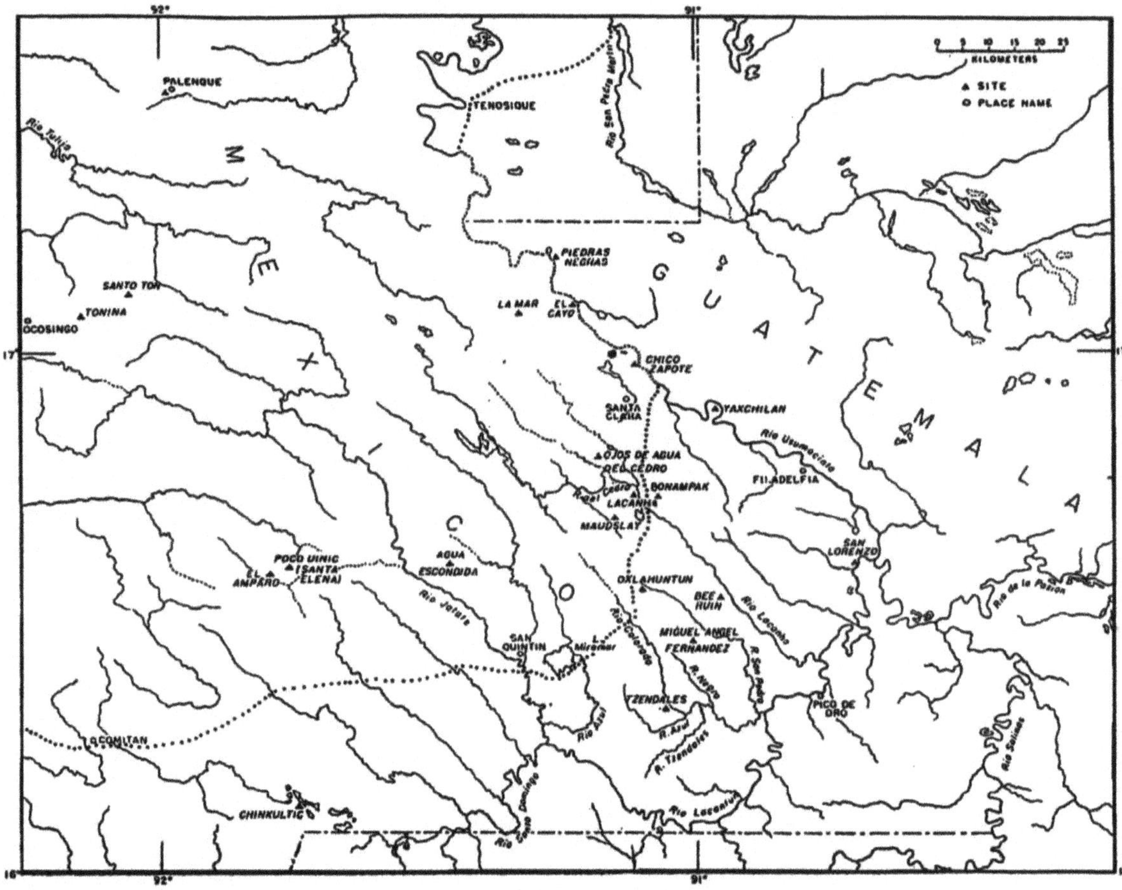

Figure 3-3 Alonso Dávila's *entrada* to the Upper Usumacinta region in 1530 (after de Vos 1988

Charnay's 1887, Maudslay's 1889 and Maler's 1901 and 1903 accounts of their own journeys add more precious information towards the identification of the communication routes in the study region. The fact that the three explorers used different routes is extremely useful to illustrate the feasibility of alternative pre-Hispanic routes, and how the competing polities of the Late Classic in the study region may have controlled them. This is in perfect accordance with Aliphat's (1994) proposition that access to these routes was a determinant factor in explaining site location.

Charnay's journey started from Montecristo (Emiliano Zapata). In this region the terrain is characterised by a meandering sluggish Usumacinta River and a coastal plain topography, it took in Charnay's times to travel from Montecristo to Tenosique four or five days by water, travelling upstream, and some twenty four hours by land. Charnay sent his luggage on canoe while he did the trip by land. On March 15, 1882 Charnay and his team set out from Tenosique to reach the ruins located some fifty miles distant on the left bank (downstream) of the Usumacinta (Yaxchilan). In order to reach Yaxchilan, a path of about five "leagues"(15 miles) had to be cleared on the opposite side of the river, and from that point a canoe had to be secured to ferry across to the other side (Charnay 1887:421-422, 424).

The journey proved to be very arduous, first crossing through marshy lowlands and later through dense forest. However, Charnay's account suggests the existence of a "road" (very likely abandoned by that time), when he wrote that they moved to the southeast towards the mountain range (Sierra Lacandon) on the Peten road (Charnay 1887, 424). Figure 3-4 shows the possible route that Charnay took en route to Yaxchilan after departing from Tenosique. This route was [re]constructed by comparing some of the natural features described by him with the modern maps. When we compare this route with Maler's later entry, it becomes apparent that Charnay's route took a considerable roundabout. This could be due to two reasons, either Charnay was "taken" by his men who wanted to "milk the cow" as long as they could or this purposeful detour was made to avoid the rapids. Whatever the case, it is possible that Charnay's men may have travelled at least in part along the ancient road to the Peten.

Charnay continues describing the difficulties of his journey and his account provides us with a very graphic description of moving over this kind of terrain (Charnay 1887:427):

We cross the Arroyo Yalchilan on the Guatemala border, not far from Locen, and leaving the Peten road, we steer to the south-east-south, on the path cleared by our men,

and encamp on the bank of the running stream in which we leave our dust-travelled limbs.
The next day we climb the range of hills which divide us from the Upper Usumacinta, and which are almost impassable for loaded animals. The sharp stones destroy the leather of our boots, and cut the mules' feet to pieces, while we are in danger of being lost down the ravines and precipices.

After crossing this mountain range Charnay and his men camp for the night, the sixth night after leaving Tenosique according to Charnay's count (note, however, that there is a discrepancy in the tallying of time when compared with Maudslay's dates). The following day they ascend another steep mountain range after which they encamp on the banks of the Chotal River, a tributary of the Usumacinta. On the eighth day Charnay arrives during the evening at the Yalchilan Pass, located on the right bank of the Usumacinta. The following day, due to the lack of a canoe to cross the river, Charnay decides to wait in Paso Yalchilan until one becomes available from the local Lacandon Indians. It is precisely on this day (the 9th according to his tally) that his crew spots the men from Maudslay's expedition which had been sent from Yaxchilan to pick up some provisions from a near by Lacandon hamlet. Charnay sent a message and some provisions to Maudslay with the request to borrow a large canoe so he can reach Yaxchilan. The canoe is sent the following day (the 21st of March according to Maudslay's entry), and after a three-hour journey downstream, Charnay finally reaches Yaxchilan (Charnay 1878:424-432).

Maudslay began his journey to Yaxchilan from Coban, capital of the Guatemalan Department of Alta Verapaz located on the highlands, hence taking a different route from Charnay's (Figure 3-4). His journey into the region begins on March 14, 1882 (a day prior to Charnay's), when he embarks on a canoe at a point known then as the *Paso Real*. García Moll and Juárez Cossío think that *Paso Real* is a place located in the close vicinity of present day Sayaxche (footnote no. 2, García Moll and Juárez Cossío 1990:31). Approximately three days later Maudslay arrived to the point where the Salinas/Chixoy and Lacantun Rivers converge to form the Usumacinta. Maudslay immediately comments on the changes in the topography that characterizes this river in its upper portion (A.P. Maudslay 1889:40):

The banks of the river here begin to lose their monotonous appearance, and for the first time since leaving the Paso Real, we caught sight of some hills in the distance. At mid-day we entered a gorge about a league in length, where the river flows between high rocky and wooded banks, and in some places the stream narrowed to a width of forty feet.

Maudslay passes through the smaller rapids that are located upriver from Yaxchilan, and although difficult to sort, they manage to pass them uneventfully despite the fact that the river was at a low level at this time of the year. Shortly after passing these rapids Maudslay and his crew camp for the night at an unidentified place. The following day Maudslay visits a *"caribal"* (a Lacandon hamlet) which must have been located in the vicinity of Paso Yalchilan, just upriver since it is this place where Maudslay's men were sent a couple of days later to pick up some provisions when they were spotted by Charnay's men. After this stop Maudslay and his crew continued their trip down-stream and camped for the night at the right hand bank of the river. The following day (March 18th) they reached Yaxchilan after a short one-hour paddle (Maudslay 1974:40-42)

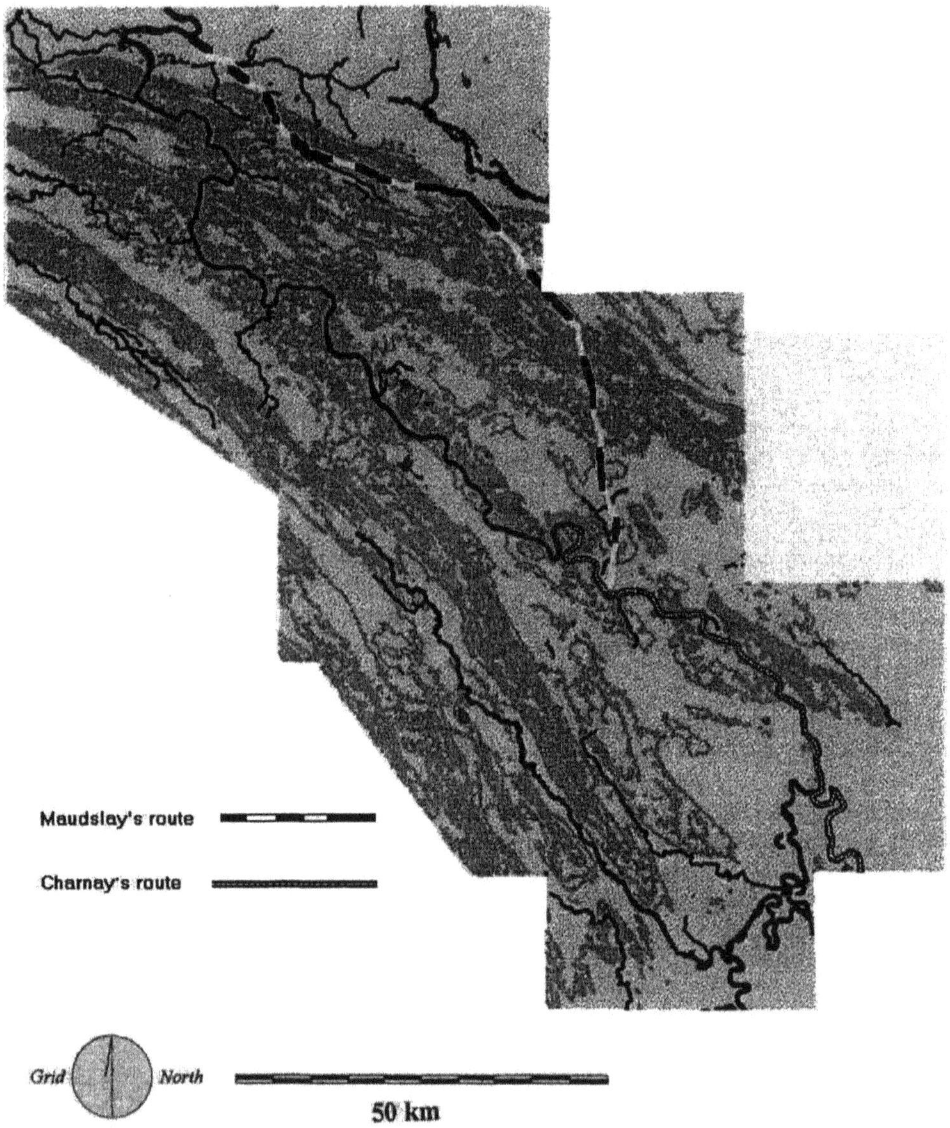

Figure 3-4 Charnay's and Maudslay's routes 1882

Maler's (1901,1903) accounts of his journey into the Usumacinta Valley are perhaps the most informative of the three. This is not only due to his graphic description of the routes, but also because his archaeological goals differed somewhat from Charnay's and Maudslay's. In this sense his endeavours were not limited to reaching a single major Maya city to record its monuments and architecture like the former, but to record as exhaustively as possible all of the inscribed monuments from the various sites in the region. In the process he also has provided us with the basic location and initial surveys of many of the very important sites in the region.

After a short stay in the settlement of Pomona (where apparently he is unaware of the existence of the ruins), Maler departs towards a settlement known as La Reforma located near the Chacamax River (which should not be confused with the important site of La Reforma-Morales located near Balancan). The trip took him two days and from this point a road known as *Camino Los Tzendales* stretched through the vast wilderness until it reached a remote settlement located along the banks of the Lacantun River. Maler mentions that the Romano lumber firm built this road. He also mentions that the condition of the road was so deplorable that not even those who built it used it in various of its initial portions preferring to take the more sinuous mountain trails (Maler 1901:9,22). It would make sense, however, that at least in part this road was laid following the ancient route that Dávila and his expedition followed.

A couple of kilometres away from the modern settlement following the *Camino Los Tzendales* and then turning towards the foothills, Maler explored a ruined city of considerable size which he called *Las Ruinas de la Reforma*. Once again following *Los Tzendales* Maler sets forth towards Chinikiha (Figure 3-5). After two leagues (approximately two-hour walk according to Maler) they turned left, this time following an abandoned *monteria* path. Another left-turn led them to the Chinikiha River reaching at this point the *camino viejo de Tenosique*, an old road, which "**passes straight through the ruined city**". Blom explored the ruins and then took the road to Tenosique for nearly two kilometers until they reached Boca del Cerro (Maler 1901:9-10 emphasis mine).

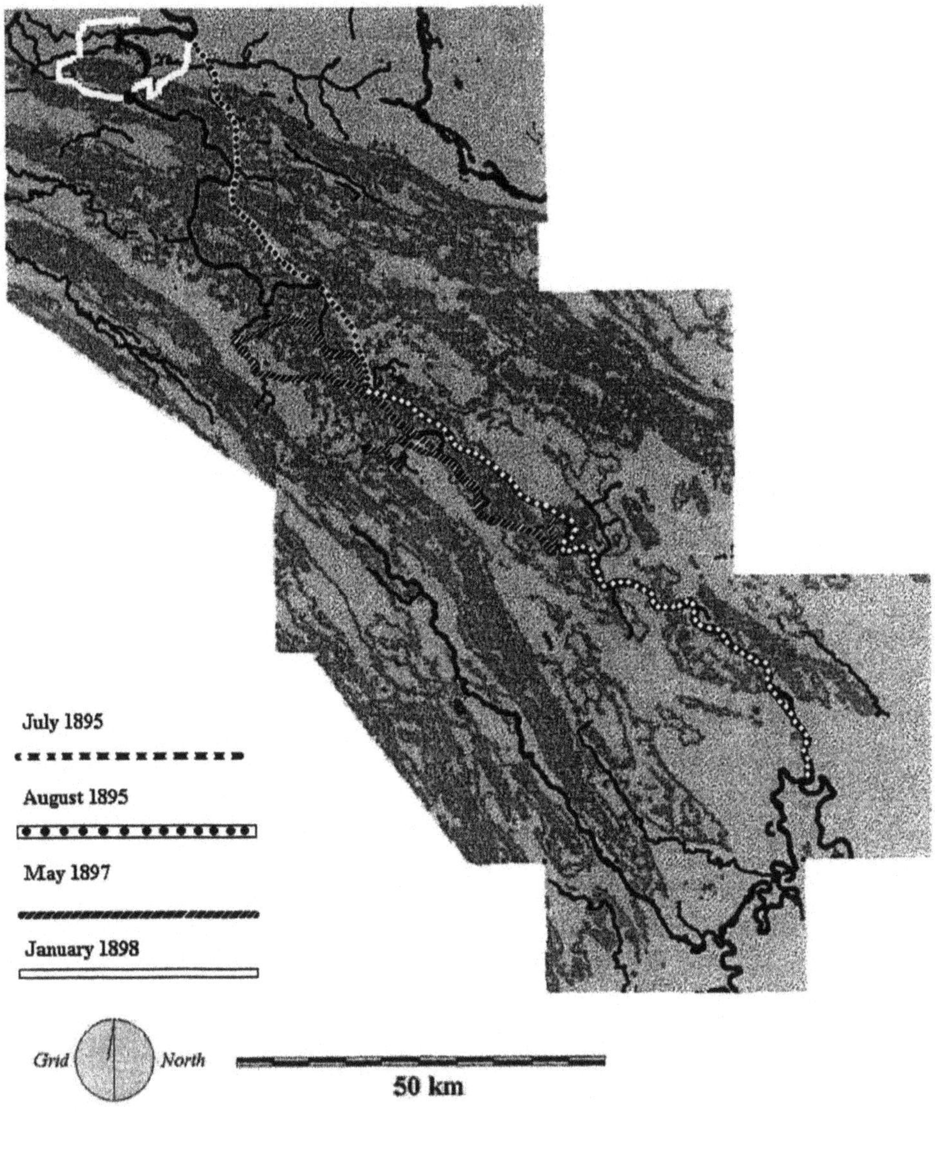

Figure 3-5 Maler's entries into the Upper Usumacinta region

In 1895 Maler takes the same way as Maudslay to get to Yaxchilan (Figure 3-4). Descending from *El Paso Real* on canoe as far as Yaxchilan, and from that point due to the perils offered by the rapids takes the forest trails all the way to Tenosique. On his way down Maler camps at El Cayo, then a *monteria*, where he learns of the existence of several important ruins including Piedras Negras and determines to prepare a later expedition.

Maler's explorations of the Usumacinta Valley begin on the 12th of August of 1895 towards the end of the rainy season, when he departs from Tenosique apparently following the same initial trail that Charnay did thirteen years before (Figure 3-4). For two "leagues" he and his crew walked through savannahs up until the small Poleva River that despite its size could represent a serious obstacle through the rainy season. After crossing the river, the trail ran mostly over low mountain ranges covered by dense vegetation that made movement difficult. The first day they camped at a *monteria* station, *Los Callejones*. The following day the journey continued through a route that took Maler through hills passing next to a sinkhole and stopping at another abandoned *monteria* station known as *Tres Champas*. The location of this place must be somewhere around the Mexico-Guatemala border, since Maler mentions that three leagues beyond this place an international boundary marker was found. On August 14th Maler and his crew arrived at Piedras Negras. It is important to note that Maler mentions that the road that goes from Tenosique to El Cayo runs by Piedras Negras (Maler 1901:41-42). At this point I think that it is safe to assume that these important *monteria* trails were laid following in large part the ancient roads. After exploring Piedras Negras for fifteen days Maler returns to Tenosique following the same road.

Maler undertakes yet another entry to continue his explorations of the sites along the Usumacinta. Departing from Tenosique on May 16th, 1897, he takes the same road as in his first entry with the intention of reaching El Cayo and from there, to continue to Yaxchilan. Maler reaches El Cayo four days later on May 20th. The island of El Cayo is surrounded by water; however, it still offered a pleasant beach. This indicates that the level of the Usumacinta was on the rise signalling the beginning of the rainy season. After exploring the ruins at El Cayo, Maler decides to visit the ruins in the vicinity so with the intention of reaching Budsilha, he and his men follow a path running downstream from El Cayo along the left bank of the Usumacinta (Figure 3-4). The initial portion of this path ran through level land avoiding the high cliffs. As he continued, however, the trail became more difficult to follow as it penetrated into mountainous terrain. Maler's intention was to reach the *monteria* camp of La Mar but as the night encroached they had to camp that night in the forest. The following morning they crossed the Budsilha River that due to the rains was deep and presented some difficulties. By noon Maler and his men had reached the *monteria* of La Mar. After resting for the rest of the day Maler explored the ruins of Budsilha on the next day. To get to them they had to cross several brooks and again the Budsilha River that at this time of the year was deep and difficult to cross. Maler estimated that the ruins of Budsilha were located at a distance of about 6 km from their camp and about a two-hour journey at the most from where the Budsilha reaches the Usumacinta (Maler 1903:83-92).

Upon his return to their camp Maler heard news of a ruined city lying at about one and a half kilometres away following a nearby brook upstream. He reached a group of ruins which he called ruins of La Mar. After exploring them, Maler and his men returned to El Cayo, but this time instead of following the route along the Usumacinta they cut across through mountain and valley (Maler 1903:93-96).

Travelling on canoe upstream along the Usumacinta is not an easy task. The distance in relation to time and effort that can be advanced is determined by the season in which the trip is done. During the rainy season the wider portions of the river offer certain ease due to the back eddies of the current that aid the canoe upstream. However this is limited to very few sections (e.g. El Cayo), and there are more cons than pros to travelling upstream during this season. For example, Maler mentions the tribulations that they had to endure while travelling in canoe from Anaite to Yaxchilan in the midst of the rainy season. These difficulties were mainly due to the fact that the poles that boatmen usually used to propel the canoes could not reach the bottom and so they were forced to reach the branches of the nearby trees to pull themselves along (Maler 1903:109).

Some of the minor rapids are also "buffered" by the higher water level; however, the downside to this is that many treacherous sharp rocks are also concealed to the inexperienced boatman. On the other hand during the dry season, when the river level is low, even the smaller rapids represent a considerable threat. In this case the only viable option is portaging; this involves carrying canoes and loads overland. But since both vessels and loads may be quite heavy, and in fact are quite difficult to transport overland, an optimal place to disembark that offers less walking distances must be chosen. All in all it seems that the dry season is best for river travel both downstream and upstream as long as one knows the location of the rapids and portages.

Whatever ease of transportation the Usumacinta River may offer, the rapids downstream from Yaxchilan represent the most formidable obstacle that the traveler encounters throughout his journey in this region. Maler was very much aware of these difficulties as he wrote (Maler 1903:96):

From Tenosique to Anaité the Usumatsintla is navigable only here and there, owing to dangerous rapids and rocky narrows, and it is thus necessary for the traveller to advance some times with pack animals and sometimes with cayucos. This renders the journey most difficult, because when one arrives at a place with pack animals he has no cayucos there, and when he comes with cayucos he has no mules; thus my progress from El Cayo to Yaxchilan was extremely difficult.

Later on Maler gives a more graphic description of the perils that the rapids pose during the dry season in December 1899 (Maler 1903:111-112):

The voyage to Yaxchilan was fraught with very great difficulties. The dangerous ***Raudales of Anaité*** *had recently claimed several human lives. Besides this, several cargoes had been sunk, the oarsmen, however, having been able to save their lives...*
...The sight of the stupendous walls of rock, which we had just passed, the din of the water, forced between rocks and rushing along at the most frightful speed, caused my men utterly to lose their heads, though they had always considered themselves excellent ***vogas.*** *They were terribly frightened and only the utmost exertion of my part had brought the trembling fellows thus far.*

After reading this description there is no doubt that the Dávila expedition mentioned above must have passed through Yaxchilan crossed the dangerous Anaite rapids and as Luján puts it, miraculously passed the rapids and continued downstream to Tenosique

Maler on his way to Yaxchilan travelled by canoe on the Usumacinta as far as the *monteria* of El Chile, located some 8 km upstream where he explored the nearby ruins located on the high left bank (Figure 3-4). From that point, due to the rapids, Maler and his party had to disembark and travel by land to another *monteria* a few kilometres upstream known as Anaite which was at the point where the dangerous rapids begin. Following a small tributary stream (called the Anaite) for three kilometres inland from the *monteria* Anaite, Maler visited a small group of ruins which he considered of little interest. He named these ruins Anaite I. From El Chile Maler also visited a big inland lake then called *La Laguna Anaite*, which is currently known as *Laguna Santa Clara*. To reach this place Maler took a "road" from El Chile that led first through a transverse ravine that soon passed through a valley. However, Maler notes that by the end of the rainy season this valley would be covered by two meters of water, thus making it impassable and making it necessary to follow the rocky paths along the mountain slope on the right hand in order to enter the road once again. Passing this valley the road branches off, the left path leads to El Chicozapote, while the right branch, which followed a southwestern direction, leads to the western end of the *Laguna Santa Clara* and runs right through a group of ruins which Maler explored and named Anaite II (Maler 1903:96-98).

From Anaite to Yaxchilan Maler travelled by canoe on the Usumacinta. However, travel was extremely difficult due to the rapids as mentioned above. The river was swollen and they virtually had to pull themselves along with tree branches: it took him and his men a day and a half to advance the more or less 13 km separating both points (Maler 1903:109).

The preceding accounts leave no doubt as to the difficulties that the physical environment posed for human travel. With such limitations it is not difficult to imagine the tremendous importance that access and control of such routes must have had for the kings of the Maya cities of the Upper Usumacinta region. The questions that still remain are why were these communication routes important? What effect did they have on the regional power balance? These and other issues will be addressed in Chapter Six, but in order to do so it is necessary to construct a working model on which to frame these questions. The following chapter deals precisely with this problem.

FOUR GEOGRAPHIC INFORMATION SYSTEMS AND ARCHAEOLOGY

PART I: THE CONCEPTS

Introduction

With the advent of the PC as an affordable, fast, and generally user-friendly computational tool, it has become possible to process in a matter of minutes, or even seconds, enormous amounts of data that otherwise would have presented a gargantuan and impractical endeavour. The complexity of the databases that archaeologists work with, which include archaeological and environmental data, made it virtually impossible to integrate these within the framework of a spatial and temporal dimension. The best one could have hoped for was the visual analysis of site distribution across the landscape (Gaffney and Stančič 1991). Within this technological innovation the development of Geographic Information Systems (GIS) is a relatively recent phenomenon, and archaeology has only just begun to tap its potential for data management and spatial analysis (Limp 1987; Kvame 1989, 1990; Allen et al 1990; Gaffney and Stančič 1991).

A GIS is in essence a spatially referenced database which allow us to store huge amounts of data and easily retrieve it, manipulate it mathematically and visualise its spatial and, to certain extent, temporal distribution (Green 1991:3; Marble 1991:9; Savage 1991:22). Hence the core of the analytic power of GIS lies in its ability to manipulate digital maps. In this context maps are no longer conceptualised as essentially pictorial renderings of the real world but as numbers. This is a major conceptual leap since according to Berry (1995), throughout our 8,000-year history of making maps these have been primarily *descriptive*, limiting their application to showing the exact location of things. Increasingly, however, maps have become *prescriptive*, containing the data necessary to implement management actions. In this process archaeologists are enabled to distil from their data (all facts e.g. ecofacts, artefact and/or site typology) information (useful facts), (Berry 1995:4-5, 11).

Berry has identified three broad categories in which all spatial modelling that is done in GIS fall: data mining, predictive modelling, and dynamic simulation (Berry 1995:1-2). The first two are the more common forms applied in archaeology. Data mining is the retrieval of specific data items or the combination of items in relation to their spatial/temporal location (e.g. all sites greater than 20 square metres, located in semiarid valleys at an altitude of 1200-1500 masl, and dating from 3000 to 1500 BC). Predictive modelling in GIS differs from simple mathematical models in the sense that it interpolates the field data into mapped variables, thus providing us with more than a hard figure, it also gives us the spatial location of the relevant variables. Predictive modelling is more frequently used in Cultural Resource Management when dealing with site prediction, for example when one wants to assess the number of sites that would be affected by the construction of a dam in a river valley.

The third category is only recently beginning to be explored in archaeological problems, e.g. Kohler's and Carr's (1996) object, oriented model. Its most promising feature is that it enables the user to interact with the spatial model by allowing him to change the variables in order to track down alternative behavioural patterns. For example through dynamic modelling we could assess what the effects of population growth were within the urban area of Tikal. What were the patterns of deforestation through time due to the massive production of lime needed for the stucco decoration of the buildings' façades? What was the recovery rate of the immediate forest during the hiatus? In other words, through the application of GIS in archaeology we are able to address not only questions like "Where is?" or "How many?" but also questions such as "What if?"

In this thesis I will apply GIS firstly to model movement over natural terrain in order to estimate cost of movement (in time/distance). This in turn will be applied for the estimation of the territorial extent of polities, and the identification communication routes in relation to the interaction between centres in the region (Central Place theory and Nearest Neighbour Analysis). Secondly GIS will be applied in the creation of models of site location through two independent approaches. The first will be through the use of a formula derived from the Gravity Model, introduced in Chapter 1, that will take into account the real distances derived from the cost surface analysis rather than Euclidean distances. The second approach will be through the application of a predictive archaeological location model, which is intended to highlight the areas that are sensitive to the presence of major archaeological sites. In both cases the intention is to predict the possible whereabouts of a series of important sites frequently mentioned in the inscriptions of the Upper Usumacinta monuments (Sak Tz'i, Hix Witz, Man, the "Knot-Site), but of yet unknown or uncertain location. Finally, with the results of these models I intend to present a reconstruction of the political geography of the region throughout the Late Classic period.

Digital Terrain Models

When making use of GIS to manipulate their data archaeologists should be aware of what "occult" processes are taking place inside the CPU black box. The above section is meant to elucidate some of the considerations that are taken (or should be taken) into account when software engineers are building the algorithms for the software packages that will enable us to model our data. Another important thing to bear in mind is that these algorithms will vary from one package to another, although the results obtained are comparable and reliable providing that we take the proper preparatory measures with our data. In this sense before choosing a package and a specific algorithm we must consider at least three factors:

(1) What are the characteristics of our raw data, e.g. scale and types of maps available, availability of Digital Terrain Models (DTM)?
(2) What sort of issues are we addressing in relation to the scale needed, e.g. artefact distribution within the site, location of optimal observation points for defensive purposes, migration routes at a regional scale, etc?
(3) What are the characteristics of the hardware readily available, e.g. processing speed, amount of RAM and hard disk space, etc.?

Modelling movement over natural terrain implies not only the estimation of the costs (in time, caloric input, or money) related to the act of moving but also controlling the nature and quality of our base digital maps. Through the use of computer-aided mapping we are now able to produce digital surfaces that include the values for the x y and z coordinates and allow us to interactively manipulate these surfaces. These models, known as Digital Terrain Models (DTM), are, as mentioned above, digital representations of the ground surface. These are known as Triangulated Irregular Network (TIN) for vector systems, and Digital Elevation Models (DEM) for raster systems (for a comprehensive discussion on vector and raster systems see Gaffney and Stančič 1991, Berry 1995, Burrough 1994).

TIN's are models that interpolate a topographic surface connecting irregularly spaced elevation points by lines and planes forming a network of triangles. In this network, the vertices of the triangles represent the elevations the lines represent degree of slope; and the polygon surface represent the aspect (azimuth) of slope. TIN's have the advantage of being less demanding in terms of computational requirements and are the best suited to represent broken landscapes. On the other hand, its main limitation is that the accuracy of the surface representation will depend on the sample size and location of the elevation points in relation to actual changes in the relief (Berry 1995: 95-98, Schneider and Robbins 1998:1—3-5). For example, interpolating from irregularly spaced points like those obtained from direct field survey will produce better results than interpolating from regularly spaced sample points like those obtained from topographic contour maps.

DEM's represent relief changes, on a continuous surface formed by columns and rows of pixels. Each pixel of the DEM holds a value representing the x and y horizontal coordinates as well as the z elevation at that location. In this sense they are more computationally demanding than TIN's, which only require the storage of the x y z values for the critical points. A common method of creating a DEM is by digitising contour lines from a topographic map on a vector platform and then converting that vector to raster format. Once in raster format a surface interpolation process is established, which will estimate elevation values at locations where there is a gap of information (e.g. between contour lines). In this process the pixel resolution chosen (which represent the real-world ground resolution) and the interpolation algorithm used will determine the amount of error built in and hence the accuracy of the model.

Determining pixel resolution can be quite a tricky process that is problem-specific depending on the scale that is needed. For example, if we need to assess mound distributions on the landscape in relation to high and low ground and the contour interval of our base map is at every metre, using a pixel resolution of 10 by 10 metres will obviously be inadequate. Likewise in a regional study involving a considerable area (e.g. 100 x 100 km) using a pixel resolution of 10 by 10 m or even 50 x 50 m would not only be too demanding in terms of computational needs, but also such resolution does not necessarily guarantees more accuracy. The latter is because a smaller pixel size means a bigger gap between contour lines that has to be interpolated thus lending itself to possible distortions especially in terrain which presents both broken (packed contour lines) and even topography (sparse contour lines) (e.g. the Upper Usumacinta region).

Schneider and Robbins (1998) present an objective method for determining pixel resolution. This method is based on the premise that the quantity of information contained in a DEM increases as the cell size decreases until a certain point when no significant increase of information occurs by decreasing pixel size. The reader is advised to refer to this source for a more detailed technical explanation (Schenider and Robbins 1998: 1—16-19).

Closely related to the previous considerations are the interpolation method applied. The most commonly used are the so-called global interpolation methods. These methods apply a polynomial function to the rasterised contours, treating the whole surface in exactly the same manner, and producing smooth surfaces that may not necessarily reflect the existing abrupt breaks on the terrain.

One can think of the interpolation process as a window that is passed across the entire surface of the raster image. As the window moves about the area it summarises the data found (elevation values) fitting a polynomial function to this data and assigning the centre of the window with this value before moving on. The final result will depend on the size of the window, which determines the number of samples taken. A large window will produce smoother surfaces while a smaller window will produce a rougher one (Schneider and Robbins 1998: 1—1-10; Berry 1995:33).

Of the global interpolation methods perhaps the most commonly used are the inverse distance, Kriging, and minimum curvature algorithms. Inverse distance uses a nearest neighbour approach. It estimates the value of the un-sampled location from the average of the values found in its vicinity. The values are weighted so as to reduce the influence of values located further away. Kriging is a more complex method that is based on the premise that variance between sample points is not uniform, with change being dependent on the distance between these points. In this sense it takes into account the trend present in the sampled data. In this method values that occur along the trend direction within the interpolation window are given a greater weight than those which are not. The minimum curvature method is based on an initial estimation of the unknown values by using the values of the sampled data. Initially it produces a rough surface which undergoes a smoothing process by means of a series of iterations that manipulate the values of the sampled data (Berry 1995:33-35).

Kellie et al. (1996) briefly discuss these three algorithms in their application to digital terrain modelling for archaeological mapping. Their main intent was to produce the most suitable DTM for the mapping of Civil War military constructions at Fort Star, Kentucky. Of the three algorithms Kellie et al. found the inverse distance (called inverse power in the authors, paper) seemed to render a more accurate depiction of the actual terrain. The authors argue, however, that despite the usefulness of the DTM in interpreting the site it does not substitute for the standard topographic and planimetric methods of mapping (Kellie et al. 1996:1-5)

Moving Over Natural Terrain

Walking, as Marble (1996) points out, although frequently seen as a simple activity represents in fact an important aspect of the relationship that develops between humans and the physical environment in which they live. "Our ability to move ourselves and our goods from place to place under various environmental conditions is a fundamental factor conditioning the human use and structuring of space" (Marble 1996:1). In the act of walking various factors should be taken into account such as: characteristics of the individual: age, body weight, health, the load to be carried and how it is carried, characteristics of the terrain (such as slope, surface conditions, vegetation), and distance to be traversed. All of these however are physical characteristics that can more or less be controlled. Other more subtle yet important factors are culturally related, or as Gorenflo and Gale put it:

...the spatial context of many human activities cannot fully be described in terms of the physical, Euclidean distance measures associated with absolute space. Studies in a number of different contexts suggest that "real" space is as much a product of economic, social, cultural, and psychological considerations as it is of physical composition. (Gorenflo and Gale 1990:242)

For one thing we need to consider that the human striding gait is far from being uniform from one cultural group to another. As Devine (1985) has shown, there is a considerable variability in human locomotion. Devine reviewed the human locomotion information available for 160 societies and proposed that in order to be able to deal with locomotion on a world wide basis, it is necessary to abstract single traits such as gait, endurance, speed, burden carrying climbing, and walking in single file (Devine 1985:552).

For example, non-industrialised societies have a higher threshold of endurance. This characteristic is rooted in socio-economic factors (the need of running in warfare, hunting, communications) and/or ideological factors (ritual, folklore, magic), and it is attested in various accounts describing daily travel of between 40 and 64 km for men and women carrying loads of up to almost 30 kg over rough terrain (Devine 1985:555-58).

Even when we are aware of the differences that moving over the natural landscape may imply among different societies we have to bear in mind that physiologically, there are limits to human bipedalism. Various tests indicate that the lower limit for walking as 0.7 m/s^{-1}, and the upper limit for walking speed is approximately 8.5 km/h^{-1} (approximately 2.4 m/s^{-1}) since at greater speeds the efficiency of walking becomes lower (Pandolf et al. 1977:579). But what is the optimal walking speed? Marble (1996) identifies, as one of the main issues to physiological researchers was the recognition of an **energy efficient behaviour**, as a fundamental feature of human motor behaviour thus:

In a freely chosen rate of activity, a rate is chosen that represents minimal energy expenditure per unit task. In the case of natural walking, the unit task traversing one metre of ground. A speed is adopted such that each metre is covered as efficiently, from the energy standpoint, as possible. (Rose and Gamble 1994:59, cited in Marble 1996:4)

In this context all things being equal, a person walking in a self-selected manner would tend to adopt an optimal walking speed of approximately 80 metres/minute or 4.8 km/hour (Marble 1996:4).

Psychological factors also have to be taken into account. On the one hand, the measure of distance will be gauged in terms of the perception of time that we assume is involved in moving across the landscape, and on the other hand, the emotional involvement that an individual or group of individuals may have in that particular act. Distance can be perceived in different manners according to the factors involved. It can be at the same time subjective and objective, depending on the factors that are implicated in our cognition of distance. Palij (1984) for example found that in the cognitive system, spatial information can take the form of a mental three-dimensional visualisation of the actual environment even in those cases where there was no visual access (Palij 1984:126). In this regard, of prime importance are those factors such as number of turns, number of stops, number of nodes, and barriers to travel. These factors have a functional relationship with the actual physical distance, including the effects of delay in the act of moving over the landscape.

In this sense, MacEachren (1980) has argued that the assumption that an individual's cognition of distance is based solely upon the objective distance between locations is doubtful. Instead he proposes that travel time should be recognised as the factor determining cognition of distance (MacEachren 1980:30). Gorenflo and Gale's (1990) model for determining patterns of settlement interaction in the southern Basin of Mexico lend support to this argument.

Emotional involvement also plays an important role in our perception of distance. Bratsfisch (1969) conducted an experiment with 65 subjects in which several distances within Stockholm and between various cities were estimated. These experiments showed that at least three factors influenced the relationship between subjective distance and emotional involvement: (1) the importance attributed to the cities, (2) the interest in the cities, and (3) the degree of knowledge about the cities (Bratfisch 1969: 244-45).

From an archaeological perspective it is impossible to directly assess aspects such as endurance, perception of travel time, or emotional involvement in estimating distance. However there is much circumstantial evidence that can be found in the historic and ethnographic sources that could be easily applied to our case (Thompson 1930; Redfield and Villa Rojas1934; Vogt 1964; Las Casas 1967; Landa 1978). Along with these written sources, a series of laboratory and field experiments have been conducted to obtain a reliable estimate of the energy expenditure involved in the act of walking under different conditions (Soule and Goldman 1972; Pandolf et al. 1977; Duggan and Haisman 1992). It is important to clarify, however, that these experiments were conducted using a total of 32 western male subjects, mainly military personal weighing an average of 71.8 kg, with average height of 1.70 m, and average age of 23.6 years old. Nevertheless the experiments showed that body weight was an important factor in determining the metabolic cost, for which it was included as a variable in the different formulae to account for variations.

A formula to calculate the energy cost of walking on 6 types of terrain at different speeds (0.66, 1.1 and 1.55 m/s) carrying three different loads (8, 20, and 30 kg) has been proposed by Soule and Goldman (1972).

$$M = \eta(W + L)[2.3 + 0.32(V - 2.5)^{1.65} + G(0.2 + 0.07\,(V - 2.5))]$$

Where:
M = metabolic rate, kcal/hr
η = terrain factor, defined as 1 for treadmill
W = body weight in kg
L = external load in kg
V = walking speed, km/hr
G = slope (grade), %

However in their empirical tests Soule and Goldman did not work with different slope gradients and so they simplified their formula to:

$$M = n(mt)\,[2.7 + 3.2\,(V - 0.7)^{1.65}]$$

Where:
M = metabolic rate, W
mt = kilograms of body + clothing + load weight
V = velocity, m/s
η = modifying coefficient for terrain other than treadmill; for treadmill n = 1.0

The types of terrain tested were blacktop road, dirt road, light brush, swamp, and loose sand. The measured values of walking at the different speeds with the three different weights were compared with predicted energy costs from treadmill walking under the same conditions (Table 4-1), and the mean of the relative ratios was taken as the terrain coefficients for each case (Table 4-2). These coefficients indicate the relative energy cost of walking across the different types of terrain with loads of 10 to 40 kg; however load, did not seem to have an independent effect for any of the terrain (Soule and Goldman 1972:706-708)

	0.66 m/s			1.1 m/s			1.55 m/s		
	8 kg	20 kg	30 kg	8 kg	20 kg	30 kg	8 kg	20 kg	30 kg
Predicted	203 W	257 W	284 W	253W	320 W	354 W	377 W	433 W	530 W
Blacktop				.85	.93	.82	1.0	.76	.92
Dirt road				1.14	1.10	1.18	1.11	.96	.98
Light br.				1.23	1.37	1.35	1.25	1.19	1.07
Heavy br.	1.94	1.58	1.57	1.75	1.62	1.40			
Swamp	2.08	1.99	1.86	2.28	1.68	1.64			
Sand	2.04	2.08	1.84	2.45	2.03	2.11			

Table 4-1. Predicted treadmill and relative ratios of energy costs of six terrains with loads
Source : Soule and Goldman (1972)

Terrain	η
Blacktop surface	1.0
Dirt road	1.1
Light brush	1.2
Heavy brush	1.5
Swampy bog	1.8
Loose sand	2.1

Table 4-2. Terrain coefficients
Source: Soule and Goldman (1972)

In a later paper Pandolf, Givoni, and Goldman (1977) revised the original formula. This revision took into account situations that range from walking at very low speeds (1.0, 0.8, 0.6, 0.4, and 0.2 m/s^{-1} carrying loads of 32, 40 and 50 kg, to standing still carrying loads of 0, 10, 30 and 50 kg. The revised predictive formula for energy expenditure was (Pandolf et al. 1977:577):

$$\mathbf{M = 1.5W + 2.0(W + L)(L/W)^2 + \eta(W + L)(1.5V^2 + 0.35VG)}$$

Where:
M = metabolic rate, watts
W = subject weight, kg
L = load carried, kg
V = speed of walking m/s^{-1}
G = grade, %
η = terrain factor

This revised formula was intended to extend the range of applications allowing an adjustment for loads as a function of body weight, thus permitting easier calculations of energy expenditure. The observations obtained by these authors confirmed that the difference between loads had a very small statistically significant level, but the differences obtained between speeds were statistically relevant obtaining the values expressed in Table 4-3 (compare with Table 4-2) (Pandolf et al. 1977: 577-78).

	Speed m/s^{-1}				
	0.2	0.4	0.6	0.8	1.0
Energy expenditure, $W*kg^{-1}$	2.15 ±0.06	2.57 ±0.06	2.93 ±0.07	3.38 ±0.08	3.97 ±0.13

Values are means ± SE;kg = body weight + external load
Table 4-3. Mean energy expenditure for slow walking with all three external loads
Source: Pandolf et al. (1977)

From the values expressed in Table 4-3 we can observe how the energy expenditure of walking at slower speeds (<0.7 m/s^{-1}) rises steadily, and according to Pandolf et al., it is due to the increasing inefficiency and lack of fluid motion. These values, however, are for walking on even terrain, the added energy expenditure involved in walking over different types of terrain appears in Graph 4-1. This graph shows that the greater the penetration allowed from the terrain the higher the energy demands (Pandolf et al. 1977:580):

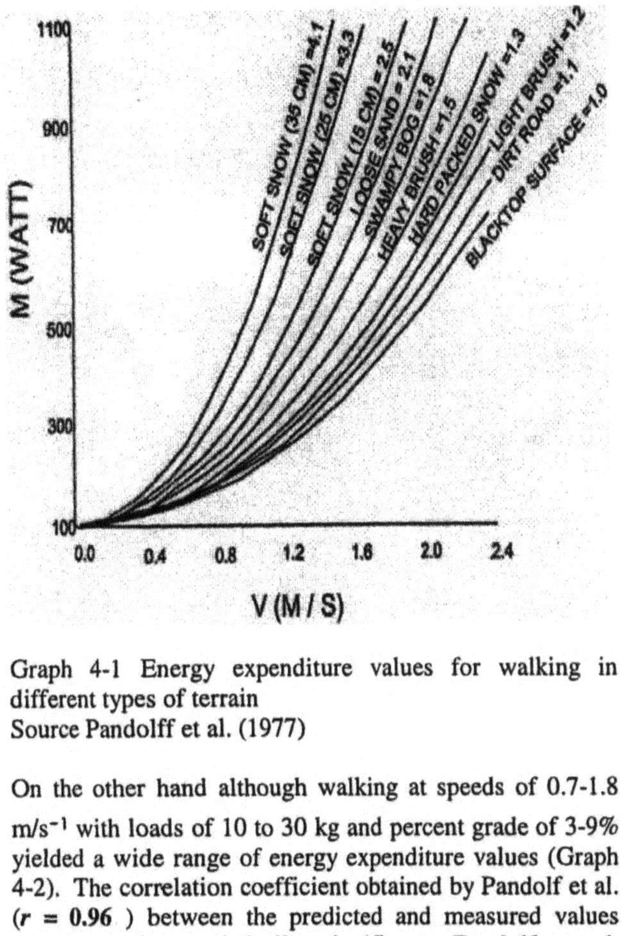

Graph 4-1 Energy expenditure values for walking in different types of terrain
Source Pandolff et al. (1977)

On the other hand although walking at speeds of 0.7-1.8 m/s^{-1} with loads of 10 to 30 kg and percent grade of 3-9% yielded a wide range of energy expenditure values (Graph 4-2). The correlation coefficient obtained by Pandolf et al. (r = **0.96**) between the predicted and measured values proves to be statistically significant (Pandolf et al. 1977:579). Duggan and Haisman (1992) did further tests of Pandolf's et al. equation, finding that the equation yielded good results when compared with the data from their own research subjects (Duggan and Haisman 1992:424).

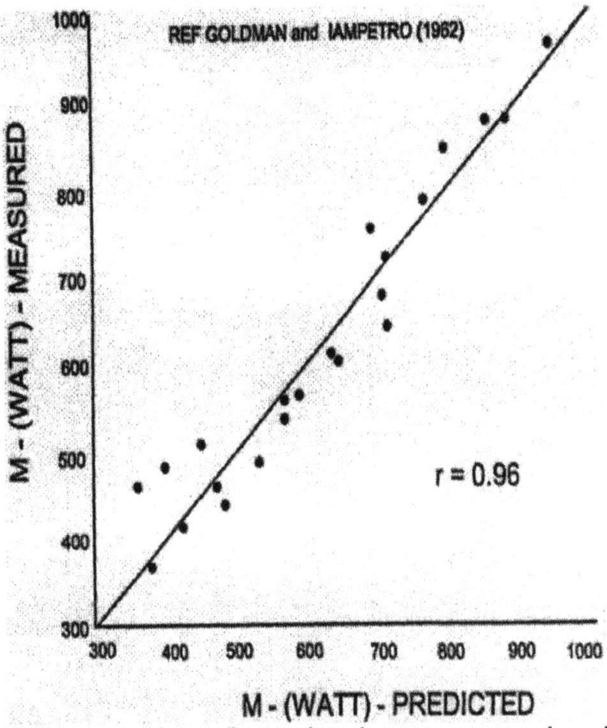

Graph 4-2 Linear Regression between expected and predicted energy expenditure values
Source Pandolff et al. (1977)

Archaeological Applications

The settlement pattern models that have been proposed for the Maya Lowlands deal in one way or another with the centralisation not only of political power, but also of goods and services (e.g. Adams and Smith 1981; Sanders 1981; Freidel 1981). In the accepted scheme polities comprise not only the main centre but also the secondary, tertiary and other subsidiary settlements included within its sphere of influence. Closely linked with this settlement structure are the various subsistence strategies in which the inhabitants of these polities engaged themselves. In essence what this means is that movement across the terrain not only implied commuting from one locale to another with a light burden, but also moving heavy loads from the periphery to the centre, or from polity to polity. This issue is particularly important if a model for trade and communications network is to be proposed.

Drennan (1984) approached this problem in terms of transport costs and the feasibility of the existence of Mesoamerican long distance trade for staple foods. Based on ethnographic observations and historic accounts, he proposes that a bearer carrying a standard load of 30 kg could cover 36 km in an 8-hour walk on even terrain. The metabolic costs of this are presented in Table 4-4 (Drennan 1984:105).

Activity	Hr/day	Cal/hr	Cal/day
Carrying 30 kg pack [2]	8	265	2120
Misc. activity [3]	2	184	368
Sitting	6	78	468
Sleeping	8	62.5	500
Total			3456

Table 4-4. Caloric requirements of overland bearers. [1]

[1] Based on Phillips (1954) and Clark and Haswell (1970:13)

[2] Not including rest periods

[3] Assumed metabolic cost equivalent to walking without a load.
Source: Drennan (1984)

Drennan also makes an estimation of the costs of water transportation taking as the basis for his calculations a "typical" dugout canoe of around 10-m long, 120 cm in diametre, with gunwales of about 20 cm above the water and a weight of around 3,000 kg. The figures for travel in and around the Peten are of approximately 20 km per day upstream and 40 km downstream per day on fairly sluggish waters. The estimated transport costs on both land and water are presented in Table 4-5 (Drennan 1984:105-107). On the other hand Gorenflo and Gale's information seems to indicate that canoe travel with light weight, was about one-third slower than travel by foot over level land, producing an average velocity for the former of 3.33 km/hr (Gorenflo and Gale 1990:244).

Medium	Days/ton-km [1]	Cal/ton-km [2]
Overland	.93	3214
Water (upstram)	.26	899
Water (downstream)	.13	449
Water (no current)	.17	588

Table 4-5. Estimated Transport costs for pre-Hiapanic Mesoamerica

[1] Days of labour required to move 1 metric ton a distance 1 km

[2] Food energy expended to move 1 metric ton a distance 1 km. Source: Drennan (1984)

Thus in the transferring food energy from one locality to another Drennan stresses the importance of comparing the gain at the destination to the food energy required moving the staples. A more detailed paper of how to calculate the cost of resource transportation is presented by Jones and Madsen (1989). Although their subject was the analysis of subsistence systems among prehistoric gathering societies of the Great Basin, their results can be applicable to farming societies as well. These authors developed a method in which the caloric returns of various natural resources were compared against caloric output when gathering these resources. Their objective was to evaluate whether a resource was likely to have been transported for long distances, thus becoming an important component of the diet. Jones and Madsen estimate the net maximum transport quantity by subtracting the cost in calories of collecting enough resources to fill a "basket load" (which can range from 7.74 to 47.67 kg), from the total caloric value of the basket load. The formulae which they obtained to measure these costs is the following (Jones and Madsen 1989:529-531):

MTD = nMTQ / (100 Cal + [1.25 Cal * x])

Where:
MTD = Maximum transport distance
nMTQ = Net maximum transport quantity
100 = the cost in calories of walking 1 km
1.25 = the cost in calories of carrying 1 kg for 1 km
x = the weight in kg of the maximum transport quantity (basket load)

In 1992 Brannan makes a revision of this model and although he recognises its applicability, he identifies four shortcomings that he then includes in his own formula. These are (Brannan 1992:56):

1) The Jones-Madsen model holds gradient and terrain type constant and therefore the caloric costs of walking constant.
2) The relationship between weight carried and the caloric cost of transporting that weight are incorrectly assessed.
3) It does not take into consideration travel to where the resource is located.

4) It provides no symbolic equation independent of specific numeric values

Similarly as in Pandolf et al. (1977) observations, the energy expenditure of walking increases as gradient also increases. Table 4-6 shows costs per kilometre at a speed of 3 km/hr for various gradients ranging from - 40% to + 40%. Brannan using MacDonalds (1961) equations obtained these values.

Percent grade	Log 10 H	Cal/km
- 40	.6452	88.4
- 35	.5838	76.8
- 30	.5615	72.2
- 25	.5389	69.2
- 20	.4771	60.0
- 15	.2976	39.6
- 10	.2635	36.6
- 5	.2985	39.8
0	.403	50.6
5	.5766	75.4
10	.7606	115.2
15	.8459	140.2
20	.9279	169.4
25	1.006	202.8
30	1.079	239.8
35	1.148	281.2
40	1.211	325.2

Table 4-6. Caloric costs per kilometre of walking at a rate of 3 km per hour at various gradients.
Source: Brannan (1992)

Brannan's modified equation permits the calculation of the final net caloric value, that is, the net Maximum Transport Quantity minus the costs of traveling from one point to another and then again to the point of origin. Another peculiarity of this equation is that it can be used for any given resource under any given terrain and gradient condition (Brannan 1992:58):

$$fnCal = \mathbf{nMTQ} - \sum_{i=1}^{n}(t_i K_i) f_i - \sum_{i=1}^{n}(t_i [W_i (L_i) + L_i]) f_i$$

Where:
t = terrain coefficient
K = caloric cost of traveling each kilometre to the resource area
L = caloric cost of traveling each kilometre back to the point of origin
W = percentage increase of caloric expenditure due to weight of transported resource
Σ: allows for variability in costs per unit of distance because of differences of gradient and/or terrain conditions.
f: denotes frequency, that is number of km values remain constant

Isotropic and Anisotropic Frictions

In GIS cost surface analysis is based on the application of cost accumulation algorithms to estimate the cumulative cost of travelling over a terrain represented in digital form as a cost-surface by incorporating the various relevant properties of the terrain. Among these considerations a very important one is direction of slope. Van Leusen (1998) decries Marble's (1996) uncritical endorsement of Pandolf's et al. 1977 formula for estimating metabolic rate. In this respect van Leusen observes that because slope calculation to obtain gradient is non-directional, no distinction can be made between down-slopes and up-slopes. Henceforth the function is limited to estimations of isotropic costs, which do not always reflect real world situations (van Leusen 1998 article presented via e-mail to the GISARCH discussion group).

In this sense cost distance analysis related to movement over natural terrain involves not only estimating the actual distance covered, but also the cost in terms of time, energy expenditure, or money that is involved in the process. To measure this we must take into account the two types of frictions present, isotropic and anisotropic frictions.

Isotropic friction is a fixed measure of resistance offered by the surface that is related to the level of ground penetration while crossing that surface (e.g. Table 4-2). This type of friction is considered to remain constant in all directions. Anisotropic friction on the other hand, is cumulative and determined by degree of slope; however, it not only has magnitude, it also has direction. This means that different friction values will be obtained depending on whether one is walking up-slope, down-slope or sideways. In order to optimally model movement over natural terrain we have to combine both isotropic and anisotropic friction costs. This characteristic makes measuring anisotropic cost a bit more elaborate since we need to combine at least three different raster images (a source, a friction and a direction image) to obtain a cost surface image. I will elaborate more on this when discussing the methodology applied in this study.

Predictive Modelling

Prediction plays a critical role in the scientific method. The mere act of proposing a hypothesis is based on the assumption that this is taking into account the observed patterns hence having predictive implications. In this sense predictive models are heuristic tools used to project known patterns or relationships into unexplored environments. Either deductive or inductive models may generate these predictions. Most predictive models in archaeology, however, are based on inductive methods. Simply stated, these inductive predictive models represent the composite patterns detected in empirical observations (Warren 1990a: 91; 1990b: 201).

More recently, regional models in archaeology have shifted interest from settlement patterns to site location ((Green 1990a; Warren 1990a, 1990b, Carmichael 1990; Williams, Limp and Briuer 1990; Marozas and Zack 1990). Although this trend by no means is signalling a waning of settlement patterns studies (e.g. Crumley and Marquardt 1990; Jackson 1990; Green 1990b; Madry and Crumley 1990). Archaeologists have developed models of site prediction through quantitative site location studies. These studies are generally based on the assumption that the physical environment (which constitutes the independent variable) has a correlation with the cultural features: the sites (the dependent variables) that can be used to predict their presence/absence (Marozas and Zack 1990).

For some archaeologists this assumption implies two premises. The first is that the settlement choices that were made by prehistoric peoples were determined, or at the very least strongly influenced, by the characteristics of the natural environment. The second is that the environmental factors that prompted the prehistoric populations into taking those choices are still preserved (at least in part) in the present landscape. From these premises it is possible to construct an empirical model for site prediction, providing that there are enough empirical observations available of the study area (Warren 1990b).

These inductive/empirical models constitute the formal devices for pattern recognition. However, we should be cautious in the weight that we assign to the physical features of the natural environment in order to avoid falling into yet another form of environmental determinism. Empirical prediction of presence of archaeological sites can be obtained through the application of probability models. These models require the presence of a mutually exclusive response to specific stimuli, in this case the presence or absence of a site (the dependent variable). This distinction is essential and constitutes the basis upon which probabilities can be calculated by tracking down the response of the dependent variable to the different stimuli given by the independent variables (Warren1990a; Kvamme 1983 in Warren 1990b).

In site prediction models some of the most commonly used environmental (independent) variables include slope, aspect, elevation, soil type, distance to water, relief, distance to basic raw materials, etc. It is very important that these models take into account in their sample data, not only those cases where there is presence of archaeological sites, but also of those where they are absent. Simply stated, the way that these models operate is by projecting from a known sample a probability surface of the study area. This surface represents graphically the projected distribution of archaeological sites.
The probability surface is obtained by training the computer to discriminate between presence/absence of sites based upon the values obtained for a number of environmental variables (Marozas and Zack 1990:165-166).

PART II: METHODOLOGY

Digital Terrain Model and Cost Surfaces

The Digital Elevation Model used in this study was generated using two different software packages, Arc-Info 7.1, in the digitising and topology building stage and Idrisi for windows Version 2 for the analytical stage. A total of 19 topographic maps at a scale of 1:50,000 were used which covered an area of 9253.8 sq. km. The maps used came from two different sources, Mexico's National Institute of Geography and Vital Statistics (INEGI), and Guatemala's Defense Department. The map quality was very good, especially for the Guatemalan section however, I did encountered some discrepancies in the location of the contour lines between the Mexican and Guatemalan sides at the Tenosique/Vertice Usumacinta section for which some rubber sheeting was deemed necessary.

For my purposes and due to the size of the study area, the contour interval used was set at every 100 metres except in those parts where changes in the relief required more detail. In these cases the contour intervals were set at 20 metres. Seven different coverage types were digitised in Arc-Info:

(1) a contour map from where the DEM was interpolated.
(2) a primary river network map for those rivers that are navigable.
(3) a minor river network which represent more of an obstacle to movement .
(4) a land use map for permanent features such as lakes and swamps.
(5) a land use map for seasonal features like *bajos*, seasonal lakes, etc.
(6) a map with the location of the primary centres.
(7) a map with the location of the known secondary centres.

All coverages were exported to Idrisi for windows version 2 (see Appendix 1 for details in the cartographic model) where they were transformed from vector to raster for manipulation and analysis. Various considerations have to be pondered when defining the pixel resolution, due to the size of the study area I tried three pixel sizes, 50, 75 and 100 metres. Of the three the 75 metres pixel resolution yielded the best results in terms of feature preservation. Idrisi applies a global algorithm that uses an exponential function to interpolate the surface, which tends to preserve the shape of steep terrain in gradually changing surfaces (Schneider and Robbins 1998: 1—8).

The DEM thus obtained (Figure 4-1) was used to generate the slope and aspect images that were in turn used in estimating the anisotropic friction values. Slope measures the angle of the terrain while aspect gives us the direction of that slope. Various field and lab observations indicate that friction increases exponentially as slope increases. Based on these and my own observations made during the Proyecto Arqueológico El Cayo 1997 field season, the anisotropic friction image for the entire region was obtained using the VARCOST module in Idrisi (see Appendix I).

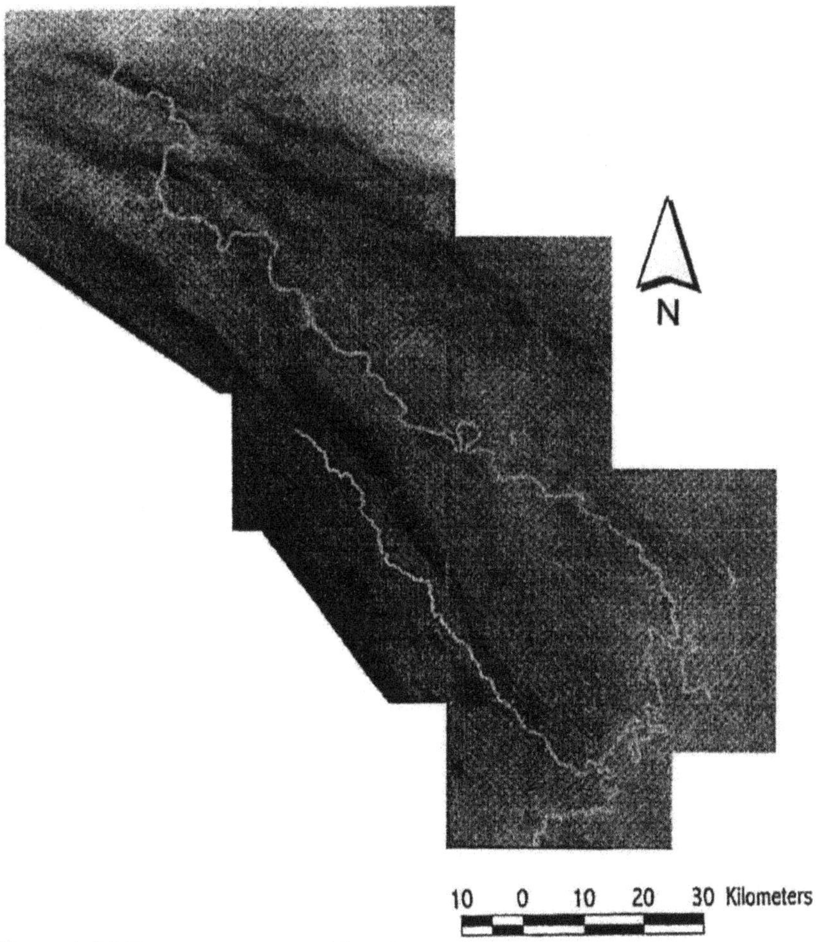

Figure 4-1 DEM generated with Idrisi for the Upper Usumacinta region

The remaining images were generated from the vector coverages using the same pixel resolution as the DEM. Overlaying the different land use images generated the isotropic friction image. Again the friction values were established on the basis of field observations.

The Usumacinta, San Pedro, Salinas, Chixoy, Lacantun, and Lacanha rivers were considered as communication routes rather than as obstacles, and except for the rapids of the Usumacinta and the small waterfalls of the Lacanha, they were assigned the minimum friction value for the downstream mode. Both isotropic and anisotropic images were overlaid to produce four different cost-surfaces. These surfaces were estimated from each of the different primary centres.

The cost-surfaces created were, downstream/dry season, upstream/ dry season, downstream/rainy season and upstream/rainy season (see Appendix 1). These in turn were used to model movement across the terrain at different times of the year.

Modelling Territorial Extent

The cost surfaces were used to model the territorial extent for the different settlements located in the study area. In the study area a total of six primary centres (Pomona, Piedras Negras, Yaxchilan, Bonampak, Lacanha, and El Chorro) with 29 subservient sites have been identified (Figure 3-1). The cost surfaces measure the cumulative cost of moving across the terrain from one point to another. In this sense if we plot a series of points in the cost image, it is possible to obtain the maximum area that can be efficiently covered from each of those locations until the whole surface is covered or it becomes too costly to transverse it.

These areas were obtained using the Idrisi's module ALLOCATE. This module creates different polygon groups by assigning the value of the initial pixel to the nearest pixel progressing until it encounters a "competing" group or runs out of surface. Figure 4-2 shows the resulting territorial extent for each of the known primary centres.

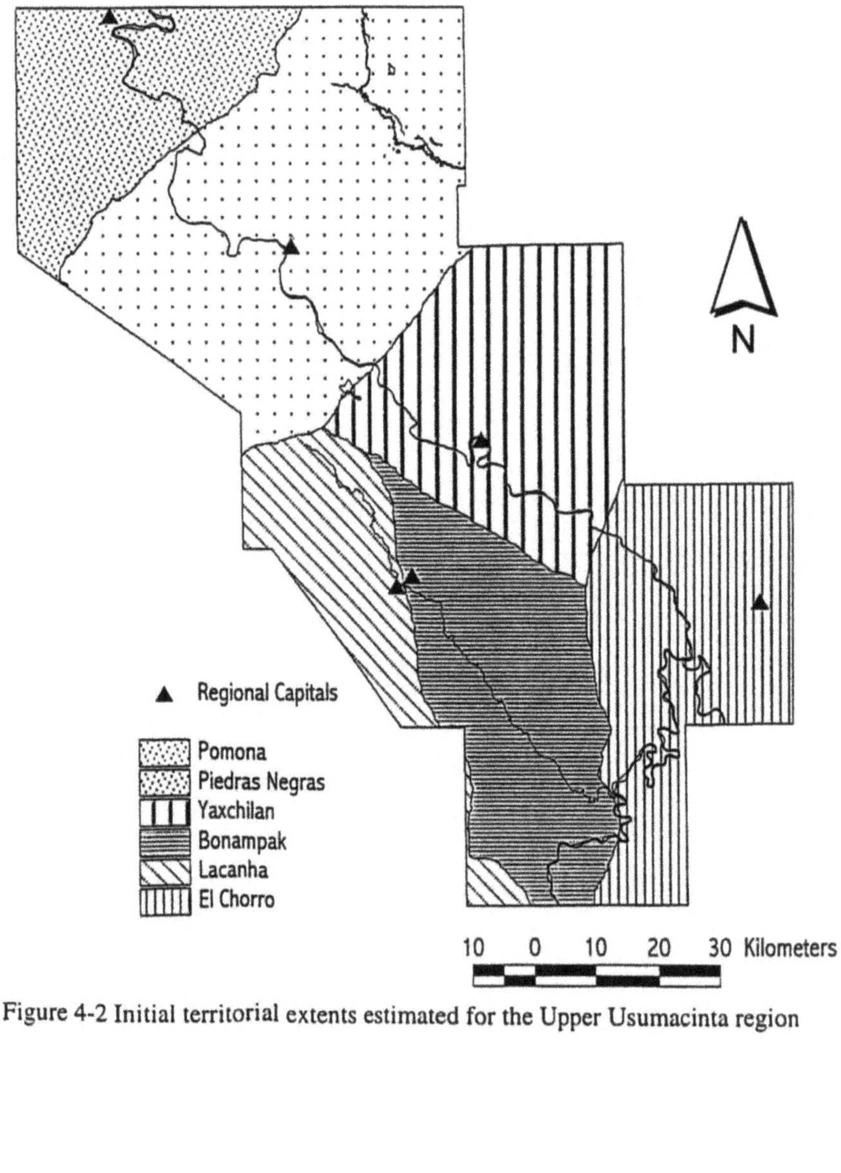

Figure 4-2 Initial territorial extents estimated for the Upper Usumacinta region

The secondary sites were overlaid to show to which primary centre they are allocated in terms of proximity. The results of this allocation procedure seem to be in accordance to the epigraphic data that has identified some of the subsidiary sites of the major polities. Although these territories match to a great extent the accepted model of site hierarchy and political predominance for the region, we have to be aware that the resulting areas are determined by the sample size, in this case six sites. To compensate for any possible bias, ALLOCATE was run once again, this time using all the sites included in Table 4-7.

Primary Centre	Subsidiary Site
Pomona	Pomona 1ª Sección
	Panhale I
	Panhale II
	Lindavista
	Chinikiha
	Sta. Margarita
	Paso El Naranjito
	Las Delicias
	San José de los Rieles
	Ojo de Agua Usumacinta
Piedras Negras	San Claudio
	Santo Tomas
	El Porvenir
	Texcoco
	La Mar
	El Cayo
	El Chile
	Anaite II
Yaxchilan	La Pasadita
	El Chicozapote
	Santa Clara
	Anaite I
	El Tornillo
Bonampak	La Cascada Lacanha
Lacanha	Nuevo Jalisco
	El Cedro
	Landeros
El Chorro	El Pato

Table 4-7 Primary centres and their subsidiary sites

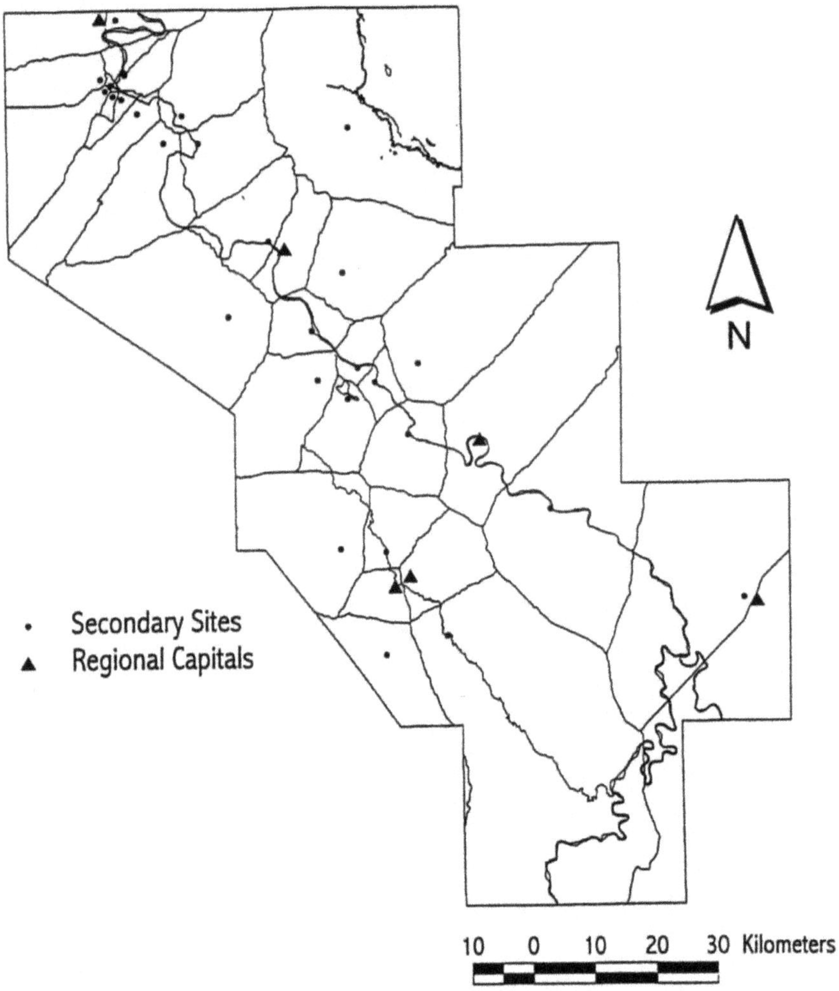

Figure 4-3 Second stage: allocation of territory per site

Figure 4-3 shows the resulting areas from this second step. The polygon identifiers were reassigned to their original primary centre identifier in order to obtain the readjusted territorial extents of the latter. The resulting territories are presented in Figure 4-4. Table 4-8 presents the total areas that can be serviced by each of these six primary centres. Again, Piedras Negras has the greatest territorial extent, followed closely by Yaxchilan. Once more these figures seem to be in accordance with the archaeological evidence. The ceramics from Chinikiha suggest that this site was a subsidiary of Palenque and hence belonged to its political sphere (Mathews personal communication 1998). If this were the case then Pomona's territorial extent would decrease. Nevertheless this does not affect the overall distribution, and we have to keep in mind that these polity boundaries were constantly changing throughout the Late Classic with the waxing and waning of the political alliances throughout the region. I shall deal with this aspect in more detail in Chapter Six.

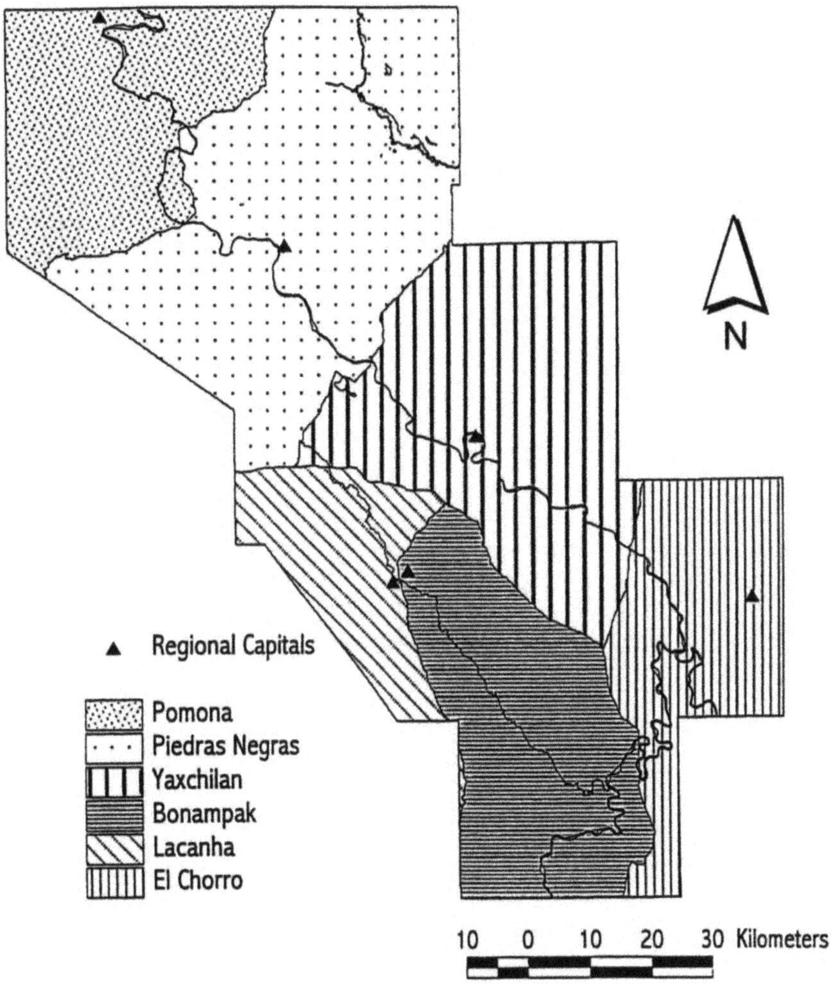

Figure 4-4 Primary centres territorial extent re-adjusted

Primary Centre	Territorial Extent (sq. km)
Pomona	1268.329
Piedras Negras	2557.220
Yaxchilan	2057.883
Bonampak	1497.193
Lacanha	724.096
El Chorro	1148.571

Table 4-8 Territorial Extent of Primary Centres

The Gravity Model

In Chapter Two, the theoretical framework of the Gravity Model was presented. Later in Chapter Three, I referred to its archaeological application in the Maya lowlands in the definition of boundaries between primary centres and the marginal development of minor centres (Dunham 1990). If we recall, the working principle of the Gravity Model is that higher levels of interaction are more likely to occur between nearest centres than with any other similar centres located further away. However as a precautionary measure size and importance of sites should be taken into consideration since larger centres tend to attract interaction from larger areas. Hence one should expect the interaction to increase proportionally to the size of the centre, and to decrease with distance. Tobler and Wineburg in 1971 presented a tantalising application of the Gravity Model in site location. In their paper *A Capadocian Speculation* the authors use this model to estimate the geographical position of several unidentified pre-Hitite towns mentioned in a series of tablets from Bronze Age Anatolia -ca. 1940 to 1740 BC- (Tobler and Wineburg 1971:). I am not aware if Tobler and Wineburg's result were substantiated or not by later archaeological research, however I felt that this approach would lend itself nicely to the Upper Usumacinta region where we have an abundance of hieroglyphic inscriptions. From these we know of the existence of a series of yet unidentified centres, Sac Tz'i, Hix Witz, Man, the "Knot-Site", Laxtunich, Buk-Tun and Wak (Mathews 1988; Palka 1996), which are mentioned in the inscriptions of Piedras Negras, Bonampak, Yaxchilan, La Mar, El Cayo, and El Chorro.

In the Gravity Model the amount of interaction between two places (marriage, alliances, royal visits, warfare, etc. (I_{ij}) is obtained by the equation:

$$I_{ij} = k\,((P_i)(P_j) / (d_{ij})^2)$$

In our case a measure of interaction may be obtained by assigning a numerical value to the different relationships denoted in the inscriptions. P_i and P_j were obtained from tallying the number of all the individual references to one site from the inscriptions of another.

The interaction between sites was measured on the basis of the type of relationship expressed in the texts (warfare, royal visits, recognition of political submission, marriage alliances, etc.). To estimate this measure, the logistics necessary to establish and maintain effective control or contact within sites was taken into consideration. For example, there are war events that are recorded as mere captures. These could have taken place anywhere between two polities. There were also wars in which the centre of one polity was attacked and burned (e.g. the Denver Panel records the burning of the "seat" of *K'ab Kan-te* of Sak Tz'i by La Mar). The latter implies actual movement of an army from one centre to the other. Marriage alliances although of prime importance for the parties involved may be established at long distances. On the other hand, the participation of foreign rulers in the domestic affairs of a different polity has different implications in terms of the effective control wielded. The values thus proposed tend to reflect spatial proximity. For example, a warfare event that records an actual attack on one of these centres will have a higher numerical value than a marriage alliance that can take place over considerable distance. There is an unavoidable element of subjectivity in assigning these values, but I think that in general terms they reflect both the actual importance of the interaction as well as distance. Table 4-9 shows the values assigned to the various types of site interaction also see Appendix 3 for a more detailed account of site interaction in the region:

Type of Interaction	Value
1. Statements of direct political subordination, e.g. "Under the aegis of…" (***u kab'ji***); "The vassal lord of…" (***yajaw***)	100
2. Statements of political alliances, e.g. royal visits; "In the presence of…" (***yichnal***)	90
3. Major war events, e.g. "Star war"; "Downed was his flint and shield" (***hub'uy u took' u pakal***)	90
4. Marriage alliances, denotes the establishment of political alliance. e.g. royal marriages	80
5. Confirmation of political ties. e.g. foreign wives participating in rituals	60
6. Taking of captive as status marker, denotes political importance of captive. "Was captured" (***chuhkaj***)	60
7. Statements that imply political peerage. "In the company of…" (***yitaaj***)	40
8. Exhibition of captives prior to sacrifice, denotes political importance of captive. "He was adorned (for sacrifice)" (***nawaj***)	30
9. Isolated occurrence of foreign emblem glyph or toponym. Denotes political interaction of unknown significance	30

Table 4-9 Numerical values for site interaction

The distances between identified sites are obviously known so the problem then was to obtain a distance for the unknown sites. For this purpose Waters derived the following equation:

$$d_{ij} = \sqrt{1/k\ [(P_i)\ (P_j)\ /\ I_{ij}]}$$

(Waters 1999 personal communication). The constant (*k*) as Tobler and Wineburg pointed out depends on each particular problem. In order to obtain this constant a data matrix with all of the sites that were referenced in the study area was constructed and a linear regression was applied (see appendix 2 for a summary of the statistics). The original data matrix included a total of 16 known sites and 9 unidentified sites, Table 4-10 lists all of the sites included in this initial matrix.

Intraregional sites	Extra-regional sites	Unidentified sites
Piedras Negras	Tikal	Sak Tz'i
Yaxchilan	Calakmul	Hix Witz
Bonampak	Palenque	Man
Pomona	Dos Pilas	"Knot-Site"
El Chorro	Tonina	Laxtunich
Lacanha	Motul de San Jose	"Site R"
La Mar		Lakamtun
El Cayo		Buktun
Nuevo Jalisco		Wak
La Pasadita		

Table 4-10 List of sites included in the initial data matrix

The distance equation was computed on a second data matrix, which this time only included the sites within the study region. The reason for this selection is that, as mentioned before, the main limitation of the Gravity Model lies in its inability to make the distinction between major centres and smaller centres. Bigger centres will tend to exercise a greater influence over a larger area than smaller ones. Hence the inclusion of major sites such as Tikal or Calakmul, or even Dos Pilas and Palenque, would certainly skew the results making them more difficult to interpret. The figures obtained ("gravity distance") after computing the equation reflected an inverse relation to distance, in other words the smaller the distance the higher the value. In order to translate these figures to actual distance the data were grouped by site and distances were calculated in two stages. First the ratio between real distance and gravity distance was obtained with the following equation:

$$rd = dab\ /\ Dab$$

Where:

dab is the gravity distance between known sites

Dab is the distance between known sites

With this ratio the distance between the known and unknown site was obtained with the following formula:

$$D_{ij} = ((dab - d_{ij})\ /\ rd) + Dab$$

Where:

d_{ij} is the gravity distance between known site and an unknown site

Of the nine unidentified sites only four offered satisfactory enough results to attempt their location: Sak Tz'i, Hix Witz, Man and the "Knot-Site". The remainder, Laxtunich, Lakamtun, Buktun, and Wak are recorded only in the inscriptions of Yaxchilan, thus they did not lend themselves to cross-referencing.

Tobler and Wineburg derived the geographic coordinates for their unknown sites from their gravity distances through a series of iterations using multidimensional scaling (Tobler and Wineburg 1971). In this study, using the distances obtained as a radius, I created a series of circles around the known centres with the hope that by overlaying these circles the possible location of the unidentified sites could be obtained. The results of these overlays are presented in Figures 4-5 to 4-8. From these images it becomes apparent that the results obtained are akin to those obtained by Dunham. Rather than pin-pointing the location of Sak-Ts'i, Hix Witz, and the "Knot-Site" the results point towards the demarcation of the breaking-point that marks the boundary between these polities. In order to evaluate the reliability of these results I applied the same process to Yaxchilan and Piedras Negras (Figures 4-9 and 4-10) to see if the definition of their boundaries coincided with the preliminary results. The results thus obtained although not a perfect match did show the potential of this method.

Figure 4-7 shows the overlay operation for Sak Tz'i from Piedras Negras, La Mar, El Cayo, Bonampak, Lacanha and Nuevo Jalisco. These results visually highlight the above-mentioned limitation of the Gravity Model. Here we can see that the gravity attraction exercised by Piedras Negras over Sak Tz'i yields a shorter distance between the two sites that is not proportional to the distances from its subsidiary sites. Also note that the distance from Bonampak covers completely that of Lacanha. The latter is due to the limited interaction recorded between the two centres, but also perhaps to the growing dominance that Bonampak had over Lacanha towards the Late Classic (A.D. 611-789).

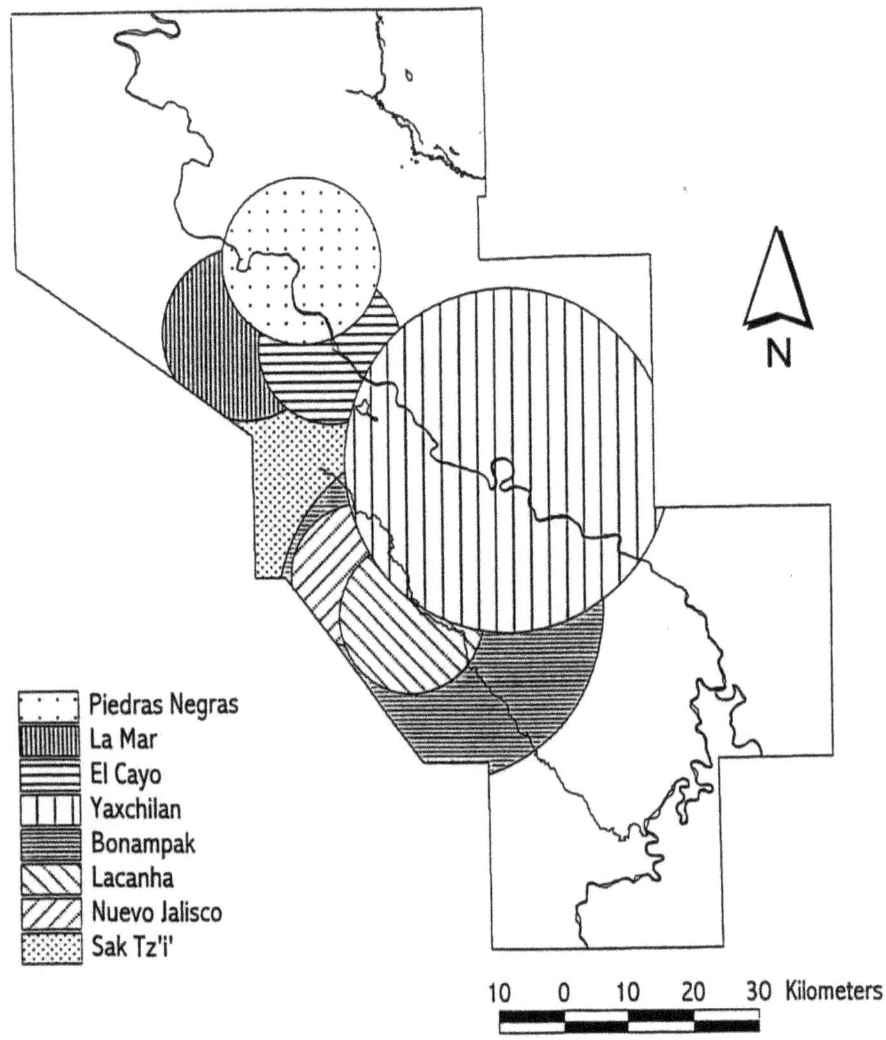

Figure 4-5 Gravity Model results for Sak Tz'i

Figure 4-6 corresponds to the overlay operation for Hix Witz obtained from Piedras Negras and Yaxchilan. Hix Witz is only mentioned in Yaxchilan and Piedras Negras; however, the intensity of the interaction recorded between these sites is sufficiently high to warrant a reliable identification while suggesting at the same time the location of the site on the Guatemalan side.

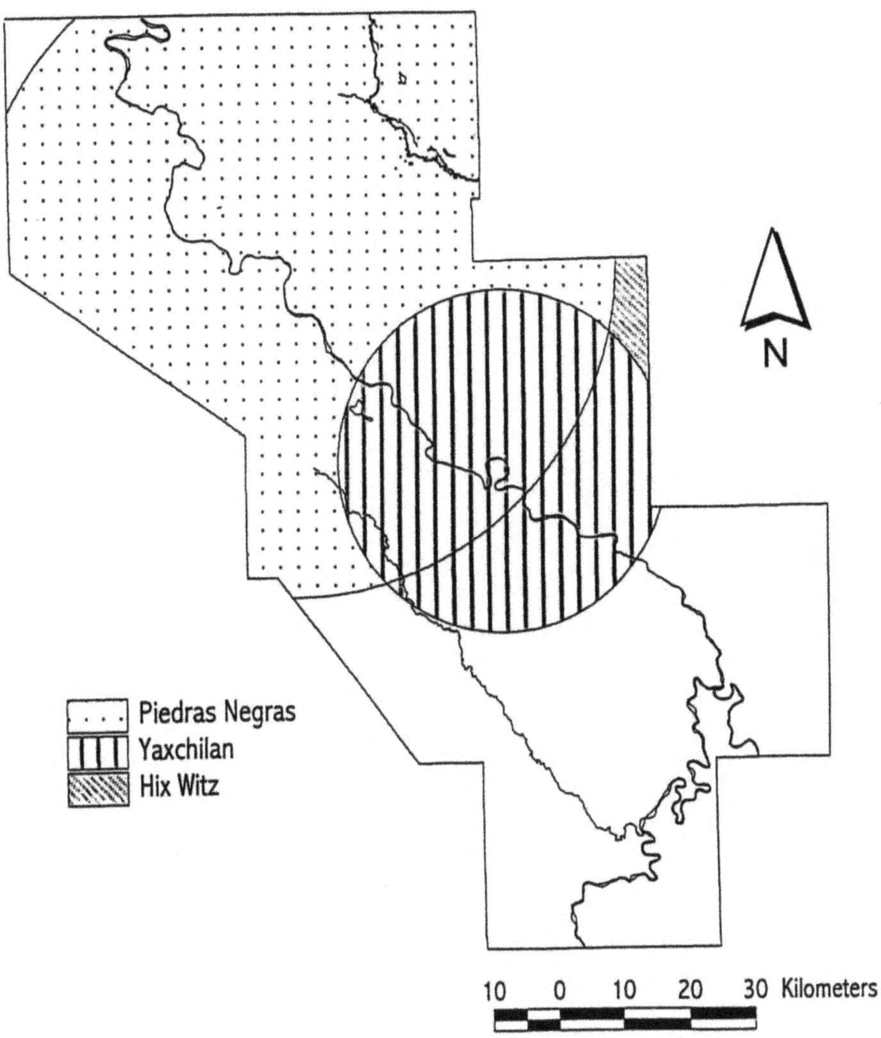

Figure 4-6 Gravity Model results for Hix Witz

Figure 4-7 corresponds to the overlay for the "Knot-Site" from Bonampak, Lacanha, and El Chorro. In this case in order to close the polygon for further analysis the estimated territorial extent of Yaxchilan was overlaid. Palka (1996) identified the emblem glyph from this site in the inscriptions of Bonampak and Lacanha. He convincingly argues that the occurrence of this emblem glyph has a geographical distribution limited to the Upper Usumacinta region, hence the site must be found here (Palka 1996:212). Dos Pilas also makes reference to this site in a war-related event (DP HSII) but since the former is located outside of the region this reference was not included in the calculation of distance. Nevertheless the reference itself is useful in providing directional trend.

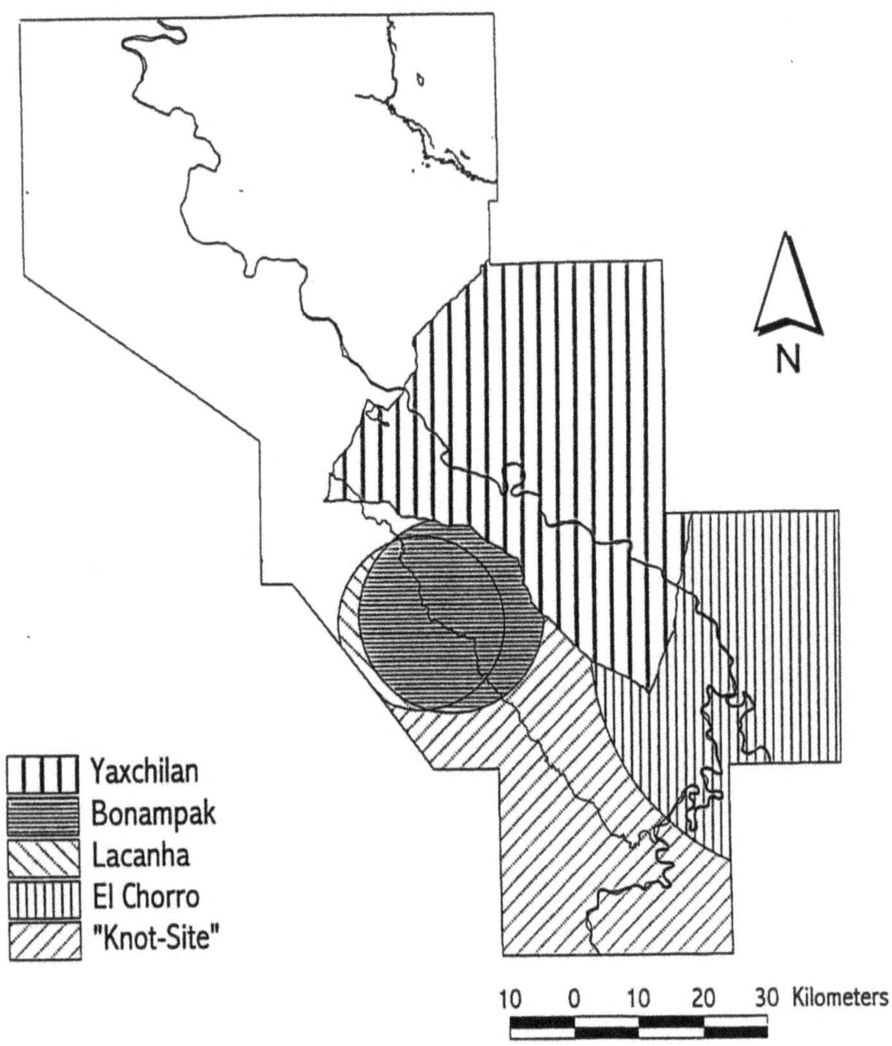

Figure 4-7 Gravity Model results for the "Knot-Site"

Figure 4-8 shows the overlay operation for the Man polity. Both Piedras Negras and Yaxchilan had a very intense interaction with this site at more or less the same time period (see Appendix 2). However as can be seen in this figure the distances obtained sets the breaking point with Man outside the region. According to this, Man is either located somewhere towards the southwest or towards the northeast. Stela 34 from El Peru contains a reference to this site, thus tipping the balance towards the latter direction. Guenter in an independent study has presented convincing data that identifies the site of La Florida with Man (Guenter 1999, personal communication). If this is the case then the apparently anomalous overlay result for Man makes perfect sense with the actual site location.

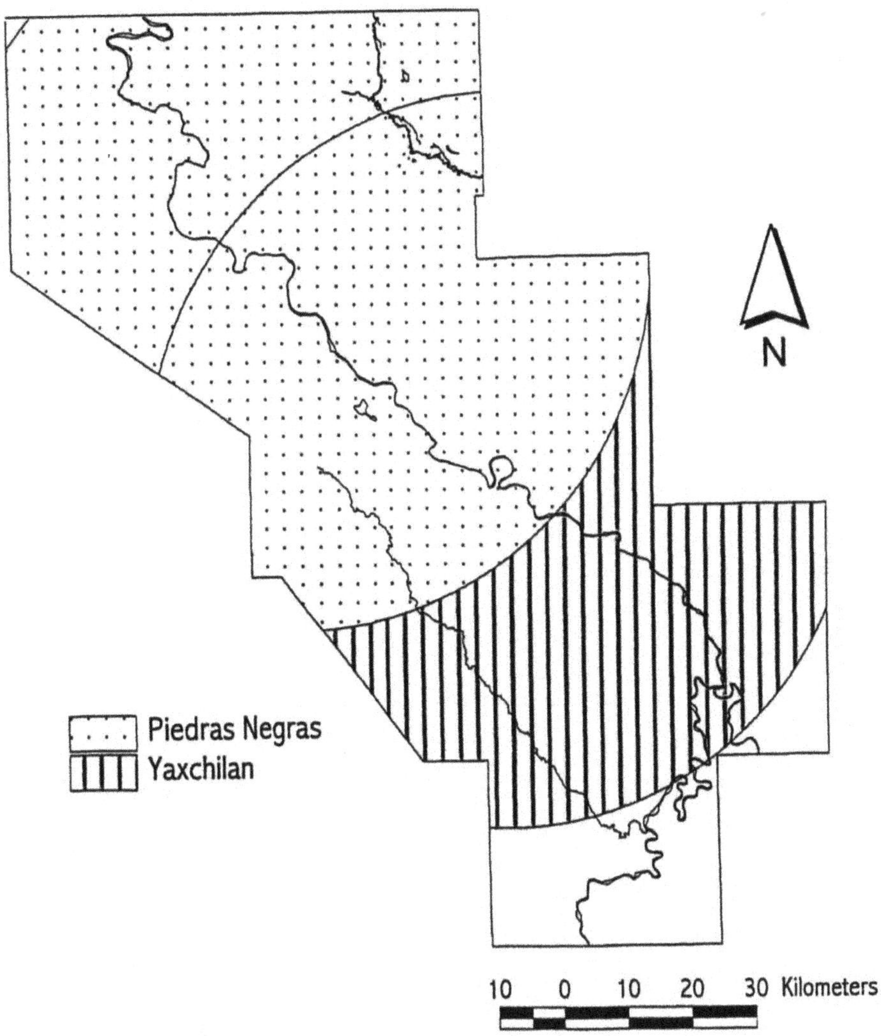

Figure 4-8 Gravity Model results for Man

In Figure 4-9a we can see the result of the test overlay for Yaxchilan. I used the gravity distances obtained for Piedras Negras, Bonampak, and Lacanha to trace the circles around these centres. In order to illustrate more accordingly these results the previously estimated territorial extent for Yaxchilan was also overlaid. Likewise Figure 4-9b shows the overlay results to locate Piedras Negras from Pomona, Yaxchilan, Bonampak and Lacanha. It becomes apparent that as the number of cross-references increases so does the reliability in the results. These tests, although not overwhelmingly convincing, do show that the use of the Gravity Model is a good exploratory approach for site location based on site interaction. It provides us with a minimum framework with which to compare our empirical data.

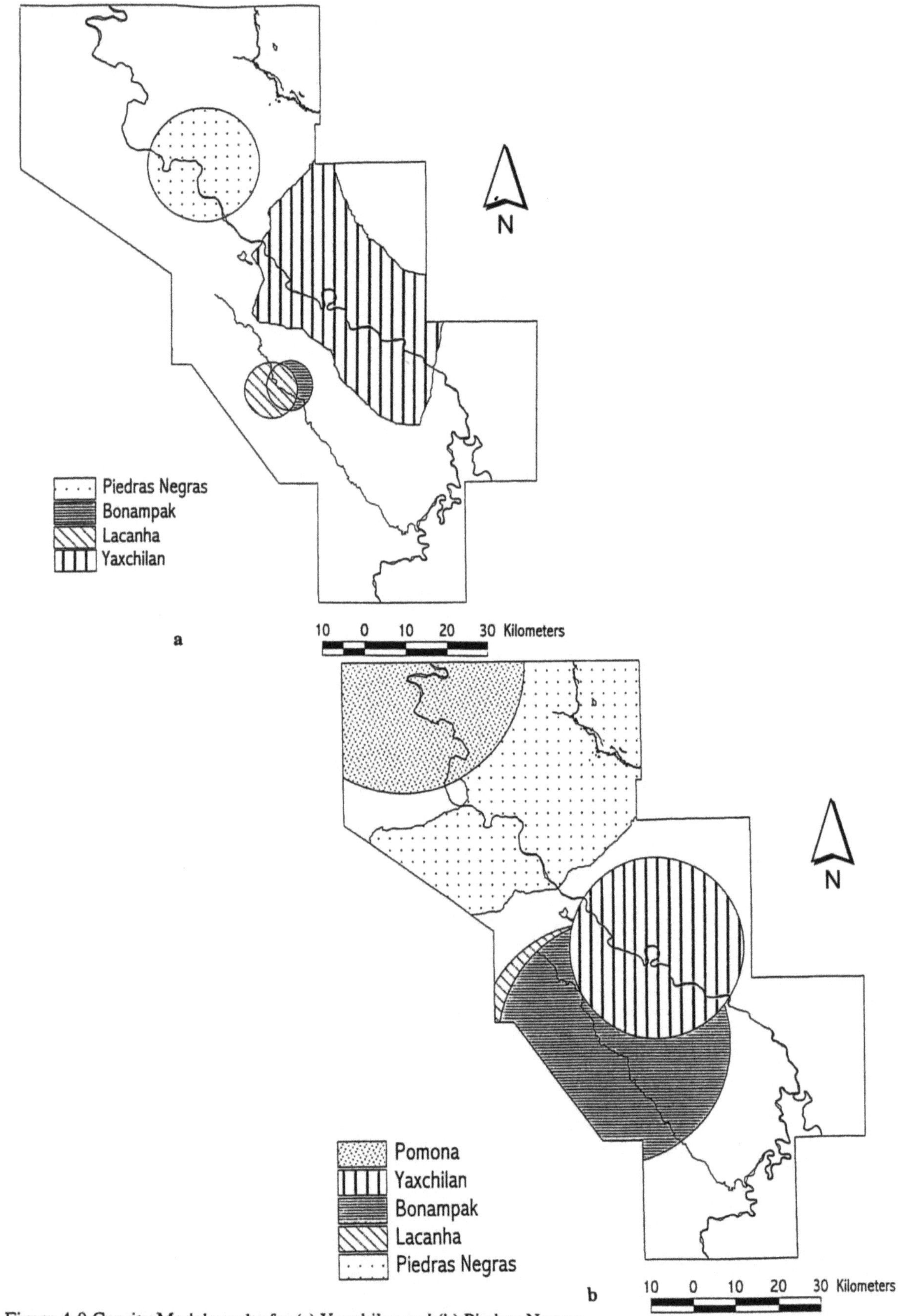

Figure 4-9 Gravity Model results for (a) Yaxchilan and (b) Piedras Negras

The polygons obtained for Sak Tz'i, Hix Witz and the "Knot-Site" were isolated and the gravity centre was obtained for each one of them in order to calculate the expected total area for these polities. To obtain the territorial extent for the complete set of the 9 regional polities I followed the same procedure as described above for Figure 4-4, that is, the secondary sites were overlaid on the initial territory (Fig 4-10a) and then the identifiers of these sites were reassigned to that of the primary centre on which they fell. The final results of this analysis are shown in Figure 4-10b.

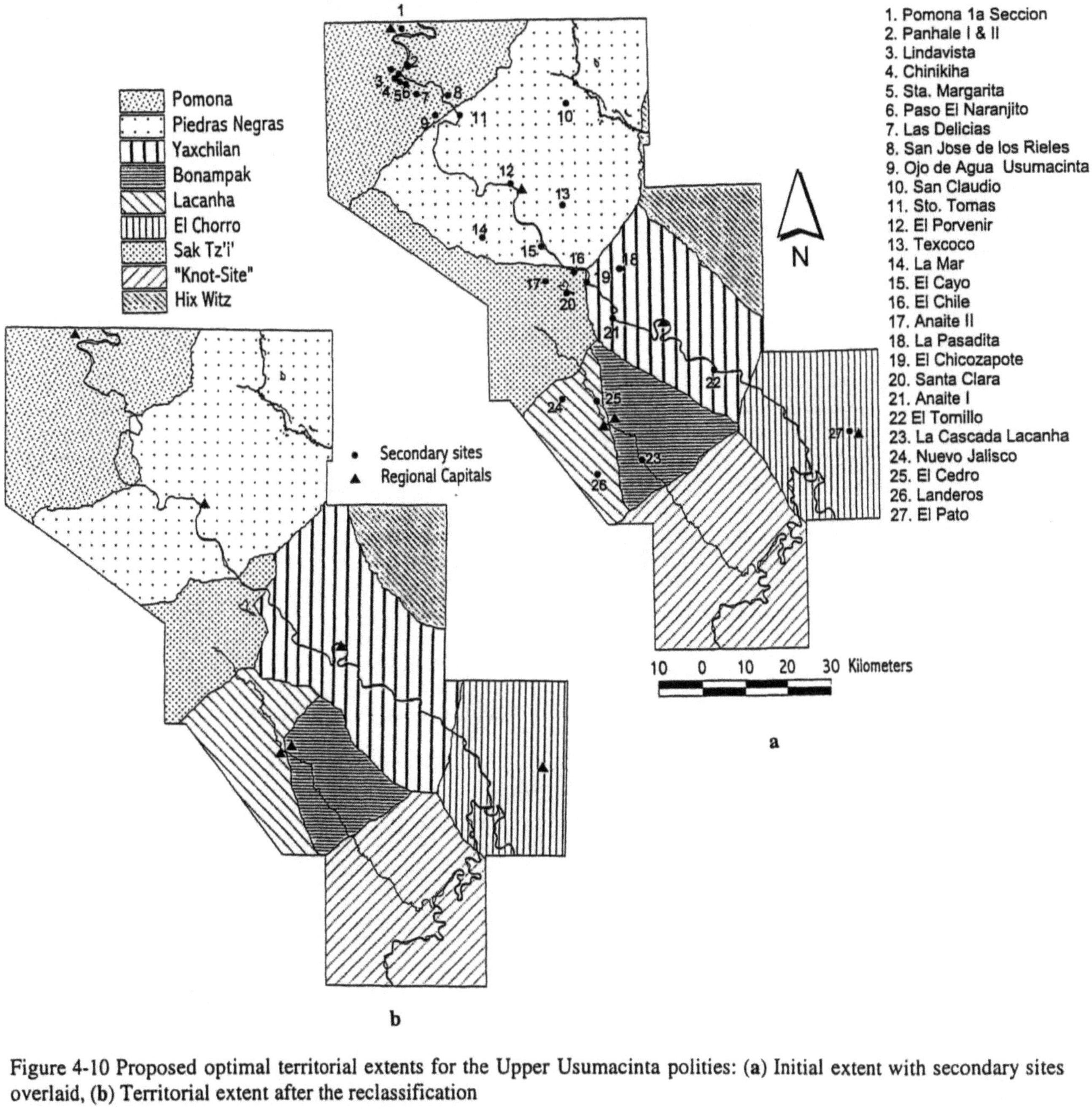

Figure 4-10 Proposed optimal territorial extents for the Upper Usumacinta polities: (a) Initial extent with secondary sites overlaid, (b) Territorial extent after the reclassification

The territorial location of the "Knot-Site" and Sak Tz'i deserves special mention. The former, although located in a largely uncharted area, occupies a central position with respect to the lower Lacanha, Lacantun, Salinas and Chixoy Rivers. This location suggests a highly strategic function as the gateway between the Upper Usumacinta and the Pasion regions. If this is the case, then one should expect to find sites positioned at critical points controlling movement in and out of the polity boundaries. In this context of particular interest is the site of La Cascada Lacanha (see Table 4-1). This site was recorded and mapped by Tovalín and Velázquez during the 1995 field season of the Proyecto Arqueológico Bonampak. La Cascada Lacanha, also known as sites 23a and 23b, is found far enough from Bonampak (a good 4 hour walk from the Acropolis) as to suggest that it is not an outlier of that centre. The site spreads on both sides of the Lacanha River situated exactly at the point where the river widens after a series of small waterfalls. Its strategic importance is substantiated by the massive size of its structures, evidence of vaulted temple buildings and sculptured monuments containing inscribed texts (Tovalín and Velázquez 1995: 41-43: Tovalín 1996). The site falls within the proposed territory for Bonampak specifically towards its southern limit, which probably marks the boundary with the Knot-Site polity. Therefore it is plausible that this site functioned in a similar way to how El Chicozapote and La Pasadita functioned for Yaxchilan, or El Cayo for Piedras Negras.

Similarly Sak Tz'i occupies a "keystone" position between the two most important polities of the region. Sak Tz'i seems to be located almost intrusively into the core of the Yaxchilan and Piedras Negras territories spreading towards the Usumacinta River and engulfing the strategically located site of El Chile, and from here extends across the Usumacinta River. From this vantage point Sak Tz'i could have easily restricted travel along the river and through the in-land routes not only for Yaxchilan and Piedras Negras, but also for Bonampak and Lacanha. In support of this idea the epigraphic evidence presents Sak Tz'i throughout most of its history as a great antagonist. It attacked and was attacked by Piedras Negras, La Mar, Yaxchilan and Bonampak (see Appendix 2). I shall discuss in more detail the impact of this spatial arrangement for the regional power balance in the following chapter. This preliminary argumentation, however, reflects the importance of having a reliable reconstruction of the territorial extent of these regional centres. With this in mind I now proceed to compare these results to those obtained through the application of the Dempster-Shafer Theory.

Predictive Modelling and Dempster-Shafer Theory

Dempster-Shafer theory along with Fuzzy Set theory and Bayesian statistics belongs to a relatively recent development of Decision Support Systems in GIS known as Uncertainty Management. This differs from the more traditional approaches in GIS where the models and databases are implied to be "perfect" thus leading us to take "hard" decisions; Uncertainty Management recognises the shortcomings of models and database. Uncertainty could be present in either the database and/or the decision rule. It includes any known or unknown error due to measurement error, instability or variability of the data set, over-abstraction or even ignorance of the relevant parameters. The strength of this approach lies not only in recognising these uncertainties, but also in providing us with the means to deal with them. In this sense we leave the realm of the traditional "hard" decisions for a series of procedures where "soft" decisions predominate. In this instance the *likelihood* of an event occurring is the parameter that is measured, which in turn will lead us to make a "hard" decision (Clark Labs 1997:9--23).

Dempster-Shafer theory is a variant of Bayesian probability theory. Both deal with the aggregation of indirect data to infer set membership of a specific phenomenon with some degree of uncertainty. However despite this common goal Dempster-Shafer departs from Bayes in terms of how the two deal with the absence of evidence. Take for example the probability of the presence of archaeological sites at specific locations, to measure this probability we establish two working hypotheses: site/non-site. While Bayes takes the absence of evidence in support of the site hypothesis as evidence in support of the non-site hypothesis Dempster-Shafer does not. This is because Dempster-Shafer theory recognizes as a working principle the concept of *ignorance*. This implies that we are aware of the incompleteness of the body of knowledge that we have of a particular phenomenon. In this context belief for a specific hypothesis does not automatically constitute the support of the negation for the alternative hypothesis. The other three very important working concepts present in Dempster-Shafer theory are *belief*, *plausibility*, and *belief interval*. The first constitutes the degree to which evidence provides support for a given hypothesis. The second constitutes the measure to which the evidence does not refute that hypothesis. Finally the belief interval constitutes the difference between belief and plausibility, and it acts as a measure of uncertainty for a specific hypothesis (Clark Labs 1997: 9--32, 9--37).

In order to obtain the above mentioned measures Dempster-Shafer requires that the working hypothesis be defined in the *frame of discernment* where they will be arranged in hierarchical combinations as shown in the following diagram:

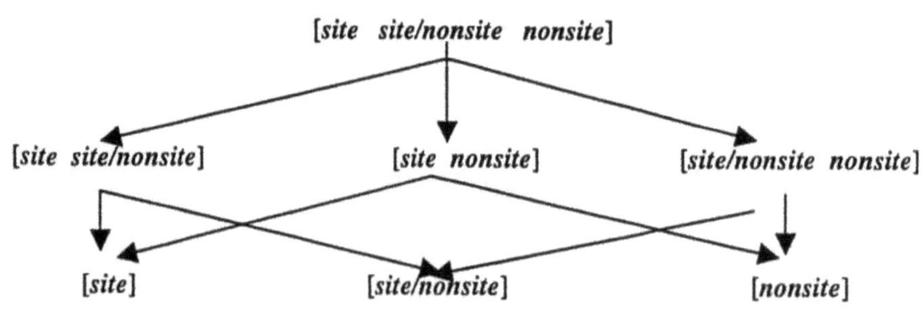

In this case *site* represents the hypothesis that a primary centre will occur at a given location, *nonsite* is the hypothesis for the absence of a primary centre at a given location. A third hypothesis, a combination of the two previous ones *--site/nonsite--* is also created. This third one is intended to exhaust all possible combination of the variables involved, thus taking into account ignorance. It frequently happens that our data may support a certain combination of variables but it is unable to distinguish further subsets of data. For example, let us say that we are studying settlement pattern development in the Tehuacan Valley using aerial photographs. We know that proximity to water, to other sites elevation, slope and aspect are some of the relevant variables for site location and are used to construct our working hypothesis. We may further wish to distinguish between *Early Preclassic* and *Middle Preclassic* sites; but on examination of the aerial photographs we find that with much difficulty we can distinguish only site from non-site. In such a case we can apply this evidence in support of the hierarchical combination *Early Preclassic/Middle Preclassic.*

Although this only offers a statement of uncertainty, this evidence can be used in the construction of the statements of belief about these hypotheses. These statements are constructed through the establishment of the Basic Probability Assignment (BPA). This represents the support that a given piece of evidence provides for one of the hypotheses but not to their proper subsets (Clark Labs 1997:9--37-38). Dempster-Shafer theory offers the advantage of being able to aggregate new data to the current state of knowledge. This should be particularly appealing to archaeology due to the additive nature of the archaeological data.

I applied the Dempster-Shafer logic to my data set in order to assess the likelihood of the proposed territorial extents for the unidentified primary centres obtained from the Gravity Model. It is important to mention that this predictive model is exclusive rather than inclusive. In other words my main concern was identifying the more likely locations for primary centres in relation to the other known centres, not the identification of all suitable locations for settlements. Therefore two lines of evidence were followed distance between primary centres, and distance from main communication routes. These hypotheses were established based on my initial observations and building upon Aliphat's (1994) proposal that the location of the main archaeological sites in the Upper Usumacinta region were strongly correlated to features of the natural landscape, particularly the presence of natural passes which constituted strategic channels of communication.

Idrisi for windows version 2 contains the module BELIEF that is used to apply Dempster-Shafer logic. This module however, requires that the probability images for each line of evidence be previously created. The probability images were created with the FUZZY module, a brief description of fuzzy sets in relation to the probability images is deemed necessary to understand the mechanics of the working hypothesis.

When grouping events, natural features artefacts, etc. into classes we very seldom have a *crisp* set on which to work. Most of the times we find it necessary to establish arbitrary boundaries between classes. However sometimes when we are dealing with continuous phenomena (e.g. distance, slope, elevation, etc.) and we need to decide which group each feature is assigned to, we find this decision a bit more problematic.

Take for example distance to water: if we are modelling site prediction with respect to distance to water and establish that the maximum distance allowed is 1 km, does this mean that potential sites located at 1.2 km will not be considered? This example illustrates a frequent uncertainty problem encountered in the decision making process. In the real world the transition from one state to another is gradual, likewise in Fuzzy sets there are no sharp boundaries between classes. Here membership to the set is measured by a fuzzy membership grade that ranges from 0.0 (non-membership) to 1.0 (full membership). In order to establish the fuzzy membership grade the values are fit into a curve (Sigmoidal or J-shaped) which requires from 2 to 4 control points that will determine its shape (Clark Labs 1997: 9--33-35).

In order to obtain the probability image for distance between sites I applied a Nearest Neighbour Analysis (see Chapter One). From this I obtained the lower and upper control points from the expected distance **rA** = 21.150 km and the actual distance **rE** = 33.6 km. The result of the Nearest Neighbour analysis (***R*** = 1.6) indicates a tendency towards dispersion possibly due to competition between centres. Thus it seems that distance away from other primary centres would support the *site* hypothesis.

The main communication routes ("roads") were obtained by establishing the least-cost-paths between the different sites in the study area (Figure 5-1), then maximum distance from these in terms of cumulative cost-friction was determined using the cost-surface image described above. Hence distance away from roads was used to support the *nonsite* hypothesis.

The results of the Dempster-Shafer analysis are presented in Figures 4-12 to 4-14. For the sake of clarity in the graphic representation I simplified the images by creating six classes with the following probability values:

0.01 – 0.04 Very low
0.05 – 0.09 Low
0.1 – 0.3 Medium-low
0.4 – 0.5 Medium
0.6 – 0.7 Medium-high
0.8 – 1.0 High

Figure 4-11 shows the degree of commitment of the data set to the *site* hypothesis. In this instance, the lighter shades of colour reflect the higher probability of the presence of a primary centre in those areas that are farther away from the existing primary centres. On the other hand the dark shades indicate a low probability of the occurrence of a primary centre. These values are in accordance with the results obtained from the Nearest Neighbour analysis and lend further support to the inter-centre competition explanation for the dispersion pattern.

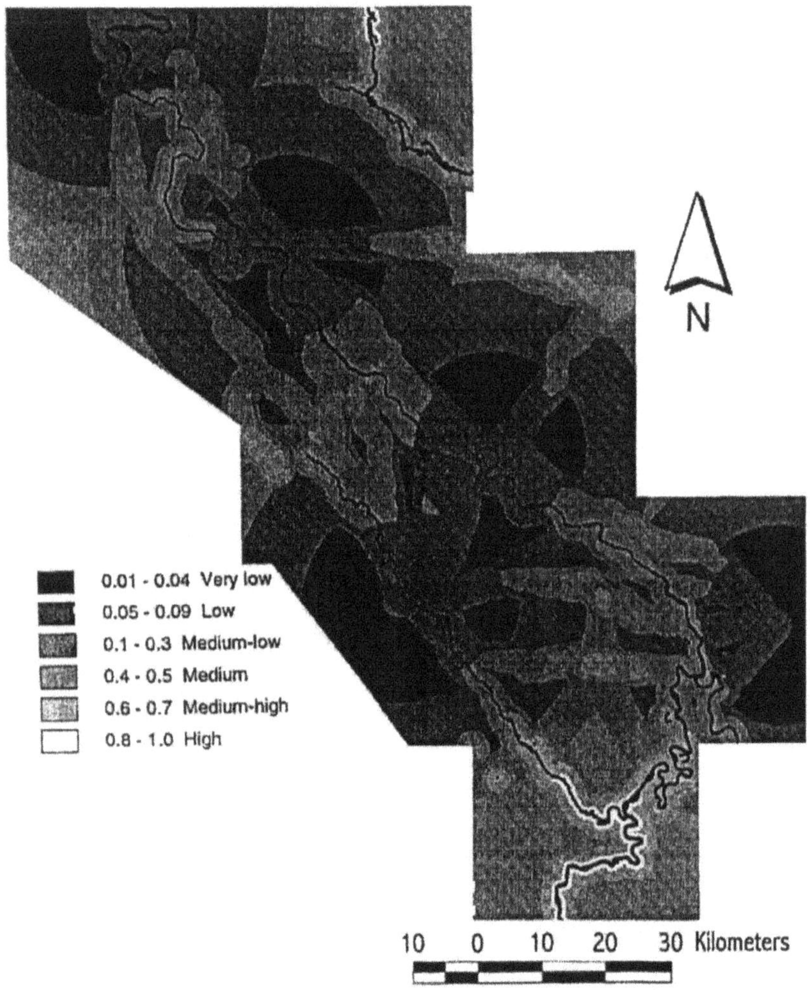

Figure 4-11 Belief results for the site hypothesis

Figure 4-12 shows the plausibility that a site can occur both at the specified distance from the other centres and within the range of the main communication routes. It is due to this that the white shaded areas, which represent a high probability value, appear thicker than in the previous image. Although this image is more inclusive in terms of the wider areas where the presence of primary centres is plausible, it does lend support to the *site* hypothesis.

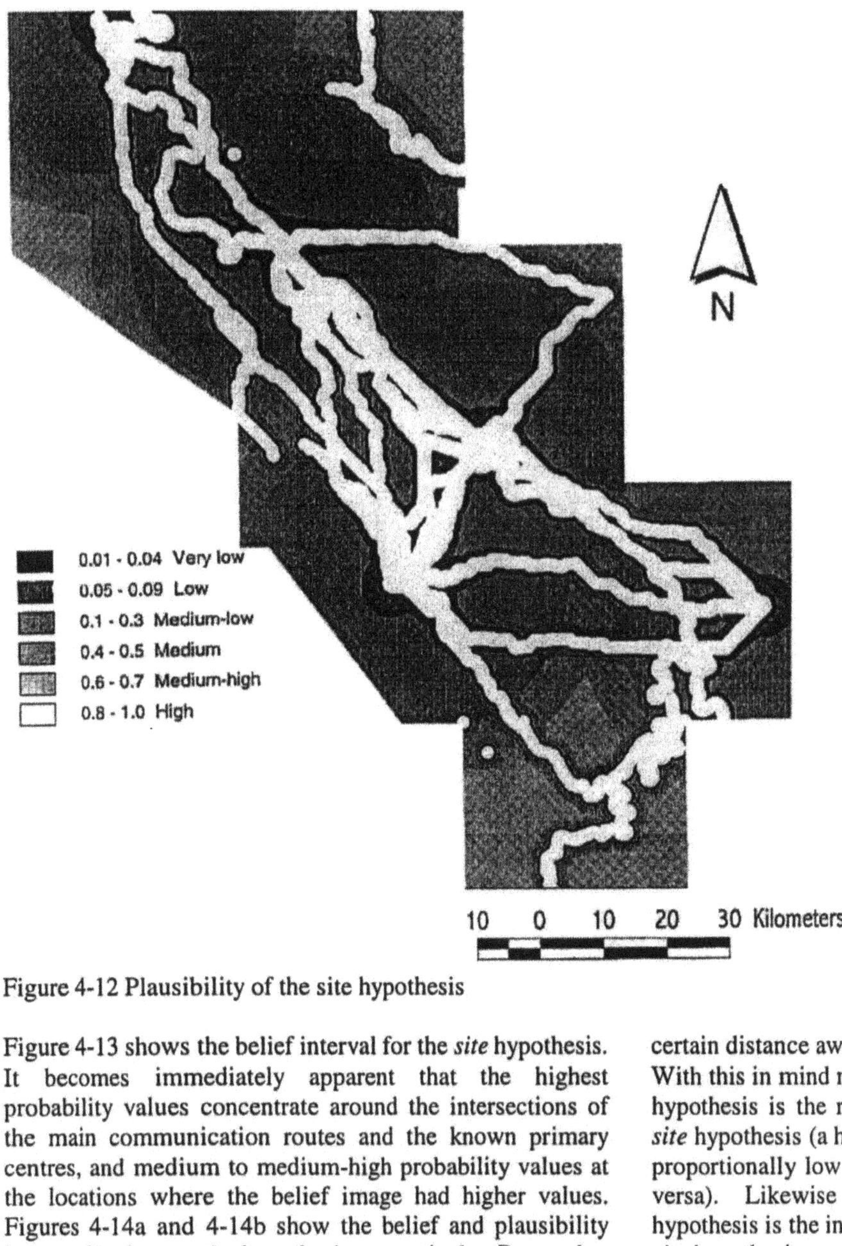

Figure 4-12 Plausibility of the site hypothesis

Figure 4-13 shows the belief interval for the *site* hypothesis. It becomes immediately apparent that the highest probability values concentrate around the intersections of the main communication routes and the known primary centres, and medium to medium-high probability values at the locations where the belief image had higher values. Figures 4-14a and 4-14b show the belief and plausibility images for the *nonsite* hypothesis respectively. Remember that this hypothesis was established to measure the probability that a primary centre **would not occur** at a certain distance away from the main communication routes. With this in mind note how the belief image for the *nonsite* hypothesis is the reverse of the plausibility image for the *site* hypothesis (a high probability value in the former has a proportionally low probability value in the latter and vice-versa). Likewise the plausibility image for the *nonsite* hypothesis is the inverted version of the belief image for the *site* hypothesis.

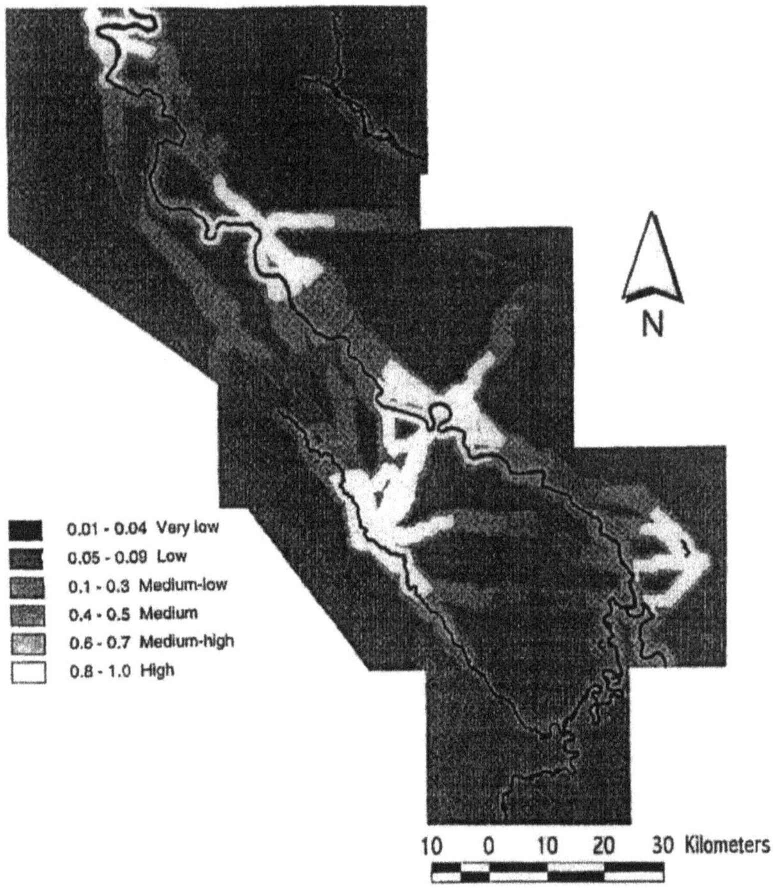

Figure 4-13 Belief interval

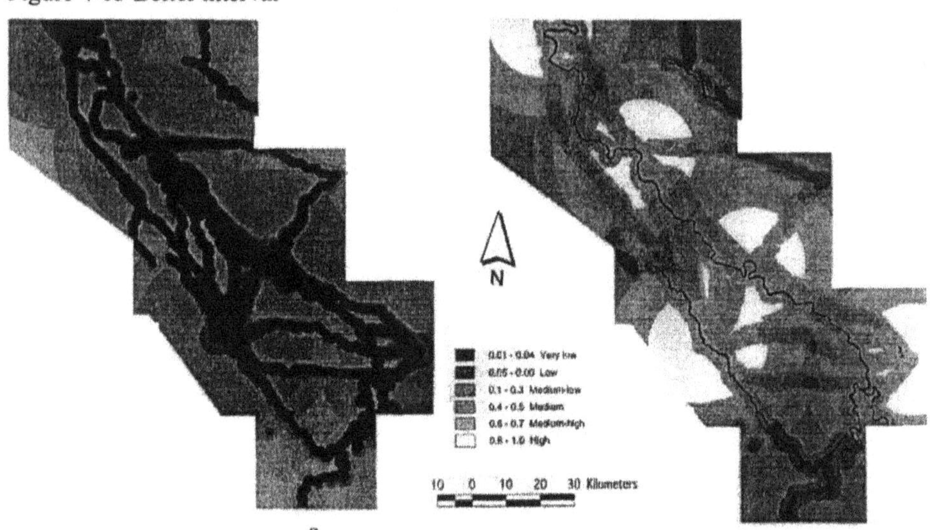

a b

Figure 4-14 Belief (a) and Plausibility (b) results for the non-site hypothesis

What can be interpreted from this set of images in terms of adding support to the likelihood of predicting the location of the unknown centres? From interpreting the belief interval image it seems that distance between primary centres, although of relevance, may not necessarily be the dominant factor for centre location. On the other hand the high probability values obtained in the belief interval along and at the intersections of the main communication routes highlight the strategic importance of these routes. These results have a better fit with the location of such subsidiary sites as La Pasadita and El Chicozapote for Yaxchilan, Texcoco and El Cayo for Piedras Negras, Panhale and Chinikiha for Pomona, and La Cascada Lacanha for Bonampak. Aliphat in his 1996 thesis pointed out that the first four of these sites were situated at critical points along these routes. In this sense these results support his arguments.

The belief interval image is of particular importance because it gave me a more serene look at the results that the initial euphoria of the belief and plausibility creates upon first inspection. Most importantly what the belief interval image shows most convincingly is the probability of potentials. In other words what this image is showing us is where the most potentially productive locations are, hence where to look for further information, and in this sense is aiding us in establishing future research designs. Does this mean that the results obtained from the belief images are negligible? Not at all, let us recall that in Dempster-Shafer theory the absence of support for one hypothesis is not evidence for the negation of that hypothesis. What the belief interval is providing us with is the level of uncertainty. This in itself is valuable enough information that can eventually lead to the substantiation of the hypothesis. The degree of commitment to the *site* hypothesis is important enough to warrant its reliability as a working model. This impression is substantiated by the medium-high probability values (0.6-0.7) that the belief interval yielded around the locations that the belief image had identified as high.

The locations assigned with a medium-high to high probability were extracted from the belief image. These were overlaid on the estimated territorial extents for the Upper Usumacinta obtained with the Gravity Model, and are presented in Figure 4-16 with the belief results in white. The correspondence between them is good, and this lends further support to the Gravity Model results while at the same time emphasising the importance of the communication routes.

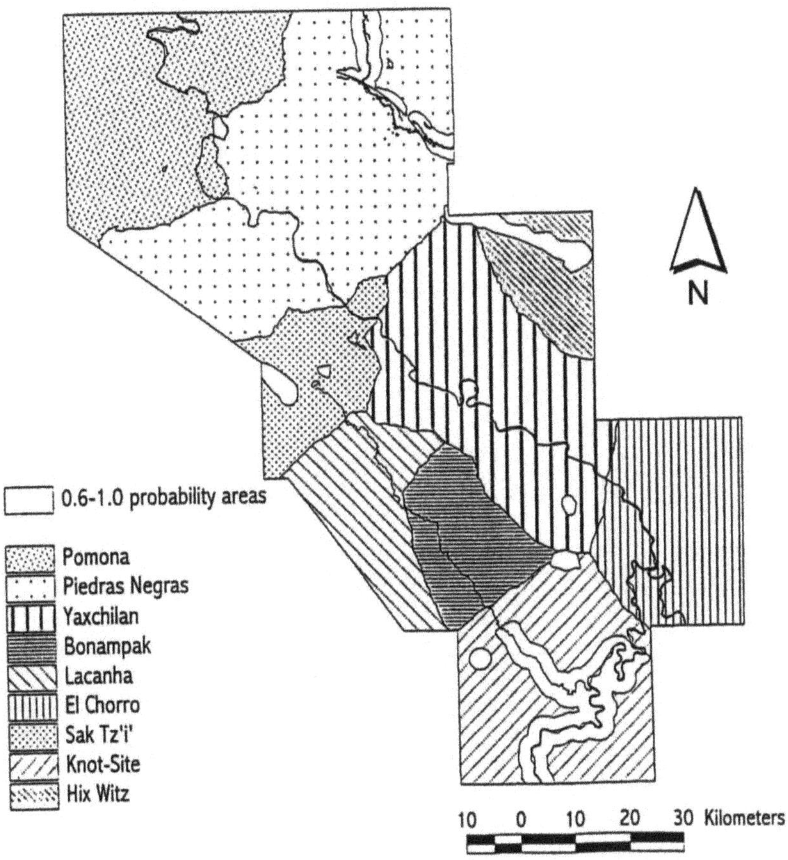

Figure 4-16 Dempster-Shafer & Gravity Model results

FIVE SITE INTERACTION AND POLITICAL GEOGRAPHY

PART I: THE SOCIO-HISTORICAL LANDSCAPE

Introduction

To the Maya elite access to written records meant the difference between glory and oblivion. It is perhaps for this reason that the Classic Maya kings went to great pains to keep a detailed account of their exploits as warriors, men of state and divine rulers. As the Classic period reached its end, a trend in the inscriptions towards a growing participation of the lesser nobility becomes apparent. This tendency has been interpreted as a reflection of the political fragmentation that was beginning to take place in the Maya lowlands (Schele and Freidel 1990). We are very fortunate to have an extensive hieroglyphic record for the Upper Usumacinta region. This hieroglyphic corpus provides us with quite a comprehensive account of the historic events that took place throughout the Late Classic.

Since Proskouriakoff's seminal work on the inscriptions of Piedras Negras (1960) and the inscriptions of Yaxchilan (1963, 1964), a series of scholars have arduously dedicated themselves to deciphering and reinterpreting the hieroglyphic texts of the region. These scholars have provided us with invaluable information for the reconstruction of the political geography of the western lowlands throughout time (Mathews and Schele 1974; Mathews 1975, 1980, 1986,1988; Schele 1976, 1986, 1991; Schele and Grube 1994, 1995; Schele and Mathews 1991; Schele, Mathews and Lounsbury 1977; Schele and Miller 1986; Schele and Freidel 1992).

When relying on the hieroglyphic record to reconstruct the structure of the political organisation we have to bear in mind that these accounts represent more the personal histories of specific rulers and their subordinates, with all the biases inherent in a self-acclaimed propaganda, rather than objective historical accounts. Mathews has frequently reminded us that when interpreting the hieroglyphic texts we have to learn how to "read between the columns". By comparing the accounts of a specific event recorded at one site with those recorded at another we are in a better position to understand the global context of events. In this work I have attempted to interpret these inscriptions in the context of the hegemonic struggle between Tikal and Calakmul that, as Martin and Grube (1995, 1998) have shown, took place during most of the Classic Maya period.

None of the decipherments presented here are mine. I rely strongly on the magnificent works of the above-mentioned scholars, particularly Schele and Grube (1994, 1995). I have also benefited enormously from long conversations with Peter Mathews, Stanley Guenter and Marc Zender. They all have generously shared with me their exciting findings and I have taken full advantage of these findings. For omissions and errors of interpretation I hold myself responsible; for any achievements I acknowledge my colleagues.

As mentioned earlier in Chapter 2, Martin and Grube have demonstrated through their analysis of specific expressions which denote subordination (*yahaw* "the underlord of...", *u-kab-hi* "it was done under the aegis of...") (Figures 2-4 and 2-5), the existence of two hegemonic powers, Calakmul and Tikal each of which wields political control over a series of otherwise autonomous kingdoms. These lesser kingdoms are linked to these superpowers by a complex network of political, military and marriage alliances that shifted from time to time altering in the process the political balance of the Upper Usumacinta and Lacanha valleys. These changes had a direct effect on the territorial extent of the different regional polities.

The principal kingdoms involved in the region were those of Piedras Negras, Yaxchilan, Bonampak and Lacanha, however these were by no means the only ones involved. Mathews in his 1988 doctoral dissertation had proposed the existence of two "buffer states", "Jaguar Hill" (currently known as Hix Witz), and the "Serpent Segment" site (currently known as Man) (Mathews 1988:379). A third important polity was also later identified, Sak Tz'i' which as I shall discuss in this section was not only a major antagonist of Piedras Negras, Yaxchilan and Bonampak, but also a critical participant and a catalysing agent for the hegemonic balance.

In Chapters Three and Four, following Aliphat's (1994) conclusion, I presented evidence that the locations of the primary centres in the region were closely linked to the locations of the main communication routes. What was left pending was the characterisation of these "roads" from the socio-cultural dimension of the landscape. The following discussion addresses this issue.

Ancient Roads and Polities

Aliphat (1994) described the distribution of sites on the landscape of the Upper Usumacinta. Of the 14 sites that he considers for the region, eight are located along the banks of the river. Two of them, Yaxchilan and Piedras Negras represent the two central places. Two other sites El Chicozapote and El Chile were established at points along the river that offer good landing beaches when portaging becomes necessary. Three more sites were established at the intermontane valley system that runs parallel to the Usumacinta, Texcoco and La Pasadita on the right-bank, and Anaite II on the left bank. This valley system offers a relatively flat and straight overland route, and all three sites are close enough to fresh sources of water, which may have been of importance to travellers (Aliphat 1994:176-182).

With these considerations Aliphat proposed that travel from Yaxchilan to Piedras Negras was mainly done along the river, except at those points already mentioned where the rapids make river travel a very dangerous endeavor and portaging becomes necessary. Upon return to the river, after a short distance El Cayo is found. El Cayo has a good landing beach and from this point it takes approximately four hours to reach Piedras Negras located downstream. On the other hand, travel from Piedras Negras to Yaxchilan would be much more complicated by river, but relatively easy by land taking the route that runs from Texcoco to La Pasadita. Aliphat roughly estimated that overland travel following the intermontane valley would take about 22 hours compared with the 17 hours that the journey from Yaxchilan to Piedras Negras takes by river (Aliphat 1994:179-181).

As Aliphat mentions the Bonampak Anticline (Sierra de la Cojolita) is cut by several faults that run perpendicular to the axis of the anticline. These faults form the gaps in the sierra that in turn provide the natural passes for the movement of goods and people between the Upper Usumacinta and Lacanaja Valleys (Aliphat 1994:31). Travel along the Lacanja valley poses fewer difficulties than along the Usumacinta. This is due mainly to the relative flatness of the terrain. With regard to water transportation the Lacanja River, although smaller than the Usumacinta, is navigable along big stretches of its course. The Lacanja is one of the various important tributaries of the bigger Lacantun River, which in turn joins the Salinas-Chixoy to form the Usumacinta, thus the water communication network for the Lacanja valley is in many ways much more complex and effective than that of the Upper Usumacinta. A few minor waterfalls, however, thwart canoe travel, but even in these cases the terrain is suitable for disembarking and a prompt portage around these barriers.

I have already briefly mentioned in Chapter 3 the nature of the communication routes in the region but without direct reference to a "formal route". Trombold defines these as "...those that show evidence of planning and purposeful construction" (Trombold 1991:3). At the present no *sacbes* similar to those found in the Yucatan peninsula or the northeastern Peten have been identified in the region; however, preliminary examination of a radar image of the region offers the possibility of the existence of small *sacbes* in those areas surrounded by *bajos*.

Considering that models derived from Central Place Theory have been frequently embraced for the characterisation of the settlement patterns of the Maya lowlands, than "roads", in this context, must have a fundamental role. Hence, in order to define the purpose and nature of formal communication routes we must take into account a couple of important considerations relevant to their functionality.

Hassig (1991) states that functional approaches to the study of roads typically adopt "economicum-transport" models to assess their utility. The main shortcoming of these models is, as he points out, that they are constructed on economic assumptions based largely on Western societies. Hence if we do not know what was considered as economically efficient by the constructing society this approach cannot properly address what the motivations for the construction of roads were (Hassig 1991:17-18).

Nevertheless, even if we cannot recover ancient concepts of economic efficiency, a fact that still remains is that roads increase the efficiency of transportation. Roads have a direct benefit in terms of cost expenditure. For example, the greater the distance a commodity has to be transported, the greater the proportion of its market price is consumed by transportation costs. However, cost of transportation should be measured in terms of energy cost rather than distance (see for example Dreannan 1984). Any change in the transportation system that increases speed, decreases travel time, hence it is time, not distance the key variable in estimating costs. Roads also play a very important social role collapsing social space by effectively drawing distant places nearer while at the same time places that are not connected are socially removed. Thus roads are selective in the ties that they create (Hassig 1991:18-20).

While considering the physical nature of roads we have once again to be cautious of not making assumptions based on our own western notion of economic efficiency. Modern roads are planned and constructed taking into account the topographic characteristics of the landscape. They tend to minimise slope changes and eliminate abrupt changes in direction even if it means sacrificing distance. On the other hand, the routing decisions for many prehistoric routes used for human foot traffic, rather than emphasising ease of gradient to avoid topographical obstacles, emphasises the directness of route. Thus these paths often pass through rugged terrain keeping a relatively straight line (Hassig 1985:32, 1991:18; Rees 1971:21-22; Sheets and Sever 1991:62).

Another very important consideration to bear in mind when analysing a communications road system is traffic volume (who is travelling), and what is being transported. The specific dimensions of the roads will dictate the magnitude of the traffic. Hassig observes that increased trade may lead to wider roads but this is not typical. Roads used for trade can be quite narrow, requiring only the necessary width to allow single line traffic. Instead he proposes that wider roads are more indicative of martial usage "The mass movement of soldiers can be crucial to the exercise of political power" (Hassig 1991:22).

Based on the premise that road width determines the distance that an army can cover, Hassig estimated that the rate of march of an army ranges from a low of 2.4 km per hour, to a high of 4 km per hour. This rate yields a day's march total distance of 19.2 to 32 km. "With these march characteristics, a single-file column for as small an army as 8,000 men, would stretch 24,000 m (15 miles), not including the accordion effect that normally occurs when marching"... "This means that, depending on the march rate, the end of the column would not begin moving until 6 to 10 hours after the march began, which would spread the army out, making the campaign longer and more costly". Hassig proposes that this problem could be resolved by increasing the number of men marching shoulder to shoulder, which of course means wider roads (Hassig 1991:23-24).

Hassig was aware, however, that there are other alternatives for moving armies expediently across the land, like stationing troops in the periphery, or by sending the army columns along several narrow roads simultaneously. On the basis of the characteristics of the Late Classic Maya political organisation and the nature of the physical landscape I think that it is safe to assume that the latter were the alternatives used by the Maya in the Upper Usumacinta/Lacanja Valleys.

The Late and Terminal Classic period in the western Maya lowlands was distinguished by what Mathews (1995) has characterised as that of "endemic warfare". In the broader context of the hegemonic political spheres created by the Classic Maya superpowers Calakmul and Tikal, the various capitals of the region constantly shifted their alliances seeking to gain regional prominence. In the process old allies became foes and vice-versa, old enemies became valued allies. In this bellicose environment the prompt dispatching of armed contingents was a matter of survival. The results obtained in the Belief Interval image described in the previous chapter highlighted the importance of locating sites at strategic points along the communication routes. In most cases these locations were coincident with the subsidiary centres of each capital. These sites are what would be considered in a hierarchical scheme derived from the titles accompanying these individuals as the sahal seats.

Various authors have amply discussed the importance of the *sahalships* (Mathews 1988; Schele 1991; Schele and Freidel 1990; Schele and Mathews 1991). As noted by these scholars, this importance increased towards the end of the Classic, and it is attested by the prominent place that these *sahaloob* occupy in the sculptural corpus. In terms of geographic location, sahal seats like El Cayo, La Mar, and El Chile for Piedras Negras, La Pasadita and El Chicozapote for Yaxchilan and La Cascada Lacanja for Bonampak occupied strategic locations in terms of access to major communication routes. From the inscriptions it becomes obvious that sahals were crucial not only to rally enough manpower to aid in the maintenance of the centre, but also to warrant the integrity of the centre's hinterland.

Sahals are frequently depicted arrayed in battle gear ready to assist their king in warfare. Some of them have the title *k'aak*, the "fire captains", the ones in charge of "putting to the torch" the helpless cities of the vanquished kings. Thus it is reasonable to assume that detachments of the kings' armies were stationed at the main sahal seats ready to be mobilised to battle upon the royal orders. Through the combination of the epigraphic data and the archaeological excavations of major defensive works throughout the Maya lowlands, scholars have generally abandoned the idea that war between the Maya was mainly at a smaller scale. These data point to the existence of armies of thousands of warriors being dispatched to battle at relatively short notice. With these considerations logistic aspects such as swift movement and close refuge have to be taken into account when analysing the war events recorded in the inscriptions.

The characteristics of the landscape in the Upper Usumacinta would have presented important challenges for the spatial integration of the various polities of the region. In order to understand the nature of these challenges and their effect on the settlement pattern, the geographic space of the region was gauged in terms of cost of travel. This factor turned out to be a good measure of proximity mainly as an indicator of spatial accessibility. Figure 5-1 shows the potential communication routes present in the study region. These routes were obtained using the least-cost-path algorithm contained in the module PATHWAY of Idrisi. Unexpectedly the results showed that movement across the natural landscape in this area proved to be more cost-efficient using the inland routes.

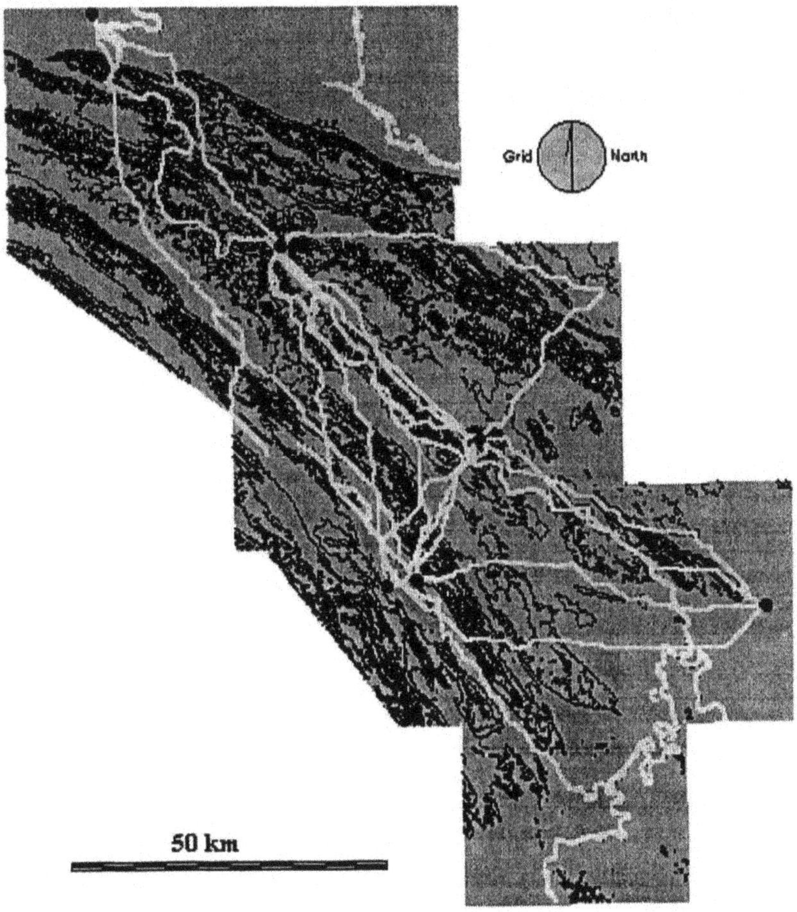

Figure 5-1 Potential communication routes in the Upper Usumacinta region

Nevertheless the "human factor" was taken into account and the river routes were obtained for those sectors where we know that the Usumacinta River *had* to be used as the main communication route in "normal" situations (i.e. from El Chile to El Cayo). The communication routes obtained with this module present a good fit to the actual known paths in the region. By comparing Figure 5-1 with Figures 3-3 and 3-4, which show Maudslay's, Charnay's and Maler's routes, you can see that they present a good match.

PART II: A RECONSTRUCTION OF THE POLITICAL HISTORY OF THE REGION

The Early Period: 8.19.0.0.0 to 9.5.2.10.6 (AD 416-537)

There is not a lot of information available for the Early Classic period in the Upper Usumacinta region. The earliest inscribed monuments are Yaxchilan's Lintels 11, 49, 37, and 35 (see Schele and Mathews 19901 Fig10.4) they cover a time period extending from 8.14.2.17.6 to 9.5.2.10.6 (AD 320 and 537). Mathews' (1988) decipherment provides us not only with the early dynastic sequence of this site, but it also gives us important information in terms of which were the important kingdoms --at least from Yaxchilan's perspective-- in the political scene of the region.

Lintel 11 (Fig 5-2) records the visit that royal emissaries from the polity of the "Diving-Bird-Site" paid to the 4th Yaxchilan king Yax Deer-Antler Chami in 8.17.13.3.8 (AD 389). I am not aware of any evidence that can assist me in the location of this polity but it must have been relevant enough to the region to merit its mention in such an important monument as Lintel 11. This site is mentioned again almost 137 years later in 9.4.11.8.16 (AD 526) on Lintel 35, which records another royal visit this time paid to the 10th Yaxchilan king K'inich-Tab-Chami II (Figure 5-2a). After this date, however, there is no more mention of the site anywhere else, so it is likely that this polity did not survive to the Late Classic.

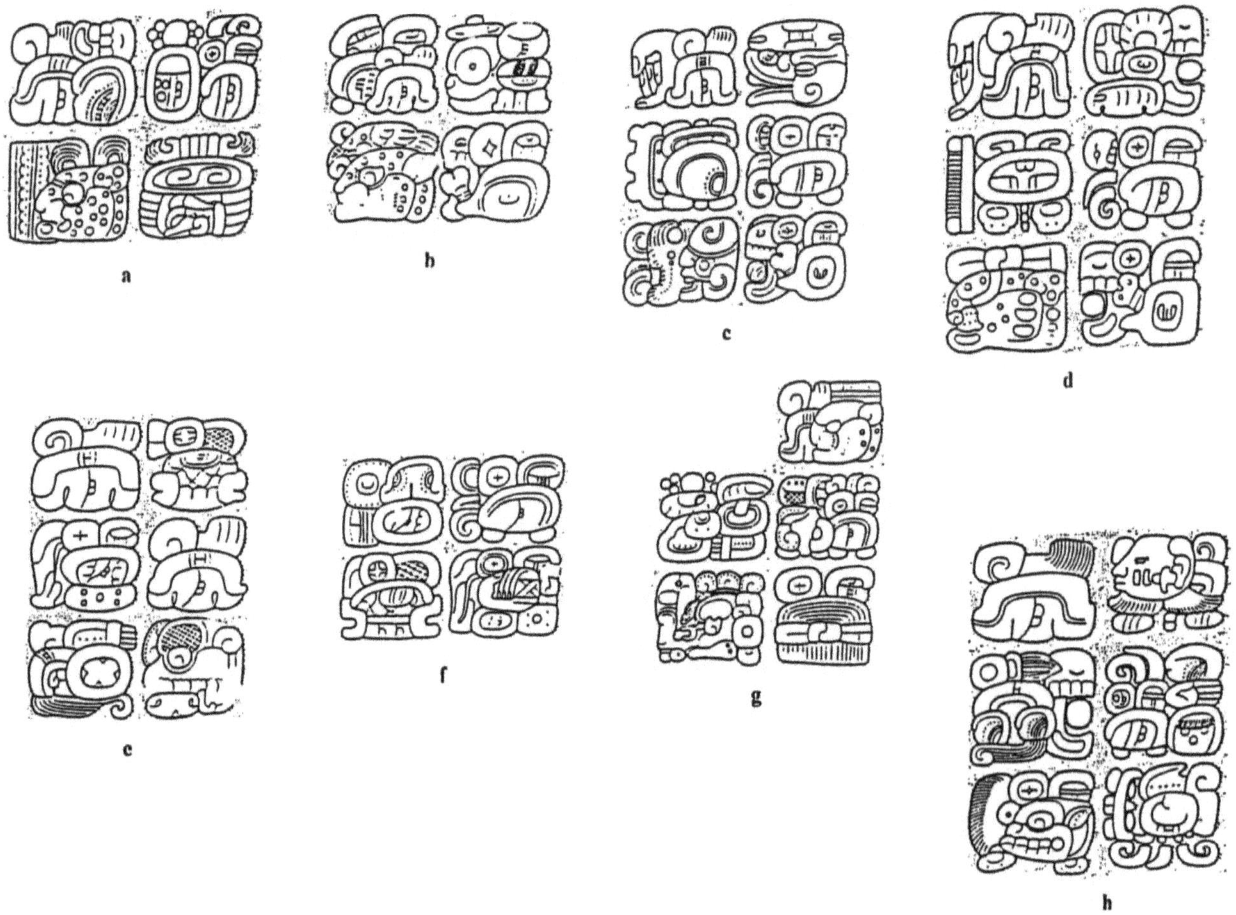

Figure 5-2 Royal visits to Yaxchilan from "The Diving-Bird site", **a** (Yax. L. 35); from Bonampak, **b**, **c**, and **d** (Yax. L. 49, 37 and 35); from Piedras Negras **e** and **f** (Yax. L. 49 and 37); from Tikal, **g** (Yax. L. 37; and from Calakmul **h** (Yax. L. 35) (Adapted from drawings by Ian Graham)

Bonampak's earliest known reference occurs in Yaxchilan in Lintel 49 (Figure 5-2b). This monument commemorates a royal visit that took place around 8.19.0.0.0 (AD 416) paid to the 6th Yaxchilan king K'inich Tab-Chami I. A second Bonampak visit was recorded in Lintel 37 (Figure 5-2c) that took place around 9.13.10.0.0 (AD 504) paid to the 9th ruler Tab-Balam I. A third Bonampak visit to the 10th Yaxchilan king K'inich Ṫab-Chami II occurred around 9.4.11.8.16 (AD 526), was recorded in Lintel 35 (Figure 5-2d). Piedras Negras is mentioned twice in Yaxchilan in Lintels 49 and 37 (Figure 5-2e and f), at ca. 9.1.0.0.0 and 9.2.0.0.0 (AD 455 and 475), also in the context of royal visits to the 7th and 8th ruler Moon Chami and Yaxun-Balam II.

In all of these instances the relationship between polities seem to be between equals. However, by 9.3.13.12.19 (AD 508) we start seeing evidence of inter-polity strife. On that date the 9th Yaxchilan ruler Tab Balam I, receives a delegation from Tikal recorded in Yaxchilan Lintel 37 (Figure 5-2g). Six years later, on 9.3.19.12.12 (AD 514) Tab Balam appears bound and in submission to the Piedras Negras king. Schele and Mathews (1991) have argued that this scene depicted on Piedras Negras Lintel 12 (Figure 5-3) represents a sort of "symbolic" captivity to the Piedras Negras king. It is not unreasonable to think that this symbolic captivity of the Yaxchilan king was the result of him "flirting" with Tikal.

Tab Balam

Yaxchilan Lord

Figure 5-3 Tab Balam I of Yaxchilan appearing as captive at Piedras Negras (PN L. 12 drawing by John Montgomery)

Twenty-three years later after the capture of Tab Balam of Yaxchilan, on 9.5.2.10.6 (AD 537), the 10th Yaxchilan king K'inich Tab Chami II receives the royal emissaries from Calakmul (Yaxchilan Lintel 35). Thus with this visit, apparently sealing his kingdom's alliance to this polity and severing any ties left with Tikal which was undergoing its hiatus period (Figure 5-2h). In this context Piedras Negras during the early period appears to have had the upper-hand, although Yaxchilan's role was certainly not negligible as suggested by the fact that it merited an official visit from Calakmul. Bonampak and Lacanha have no evidence of having received envoys from Calakmul or Tikal, and their alliance to the former seems to have been mediated by Yaxchilan and/or Piedras Negras. Figure 5-4 shows the probable territorial extent of these polities during this period.

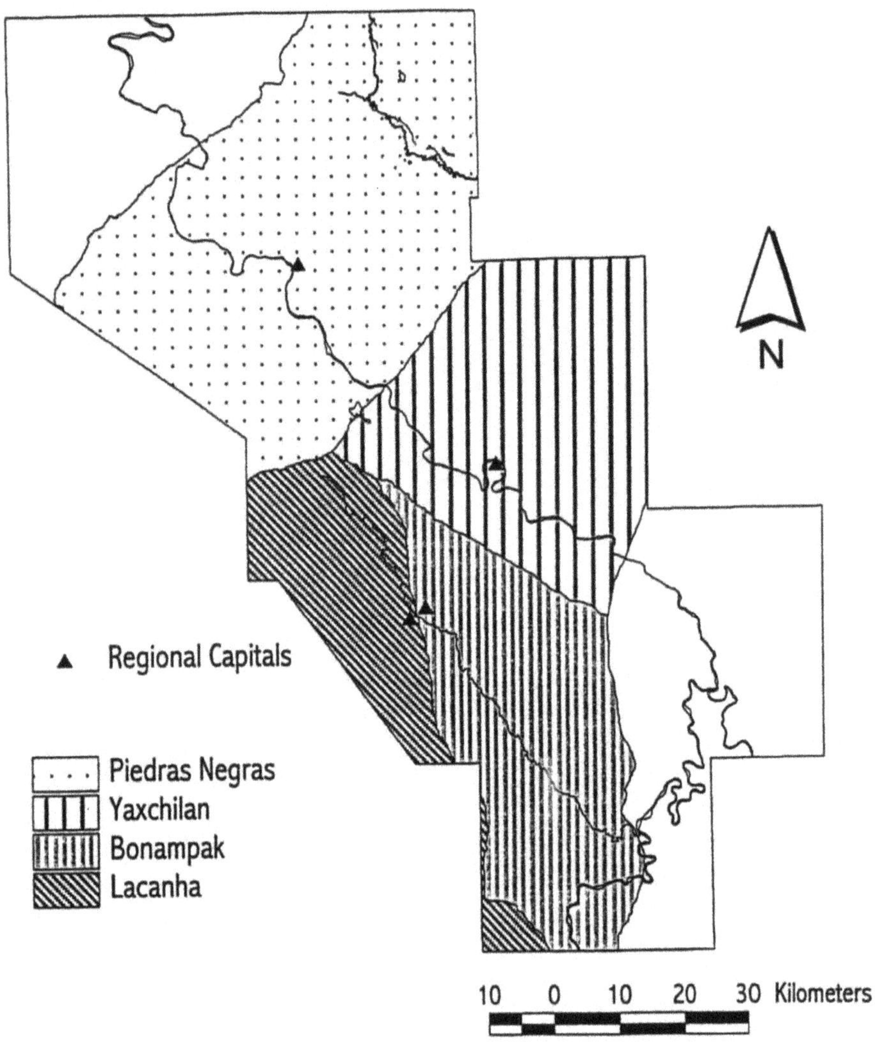

Figure 5-4 Suggested territorial extents of the Upper Usumacinta kingdoms at AD 416-537

The Early Palenque Wars: 9.8.5.13.8 to 9.9.11.12.3 (AD 599-624)

The recording of the arrival of the Tikal king Nu-Bak-Chak in AD 659 at Palenque (Temple of the Inscriptions West Panel; House C, Hieroglyphic Stairway) may indicate that these two kingdoms were long-time allies. In this context it would make perfect sense that Calakmul, either directly or through its regional allies would have waged war on Palenque and its allies. This period marks the beginning of the increasing extra-regional influence of Calakmul and Tikal that would result in a constant change of the territorial boundaries. In 9.8.5.13.8 (AD 599) Calakmul attacks Palenque probably via Pomona. Pomona due to its strategic location at the entrance of the Tabasco coastal plain was, as we shall see below, a polity strongly fought over by Palenque and Piedras Negras. It is then possible that Pomona may have belonged to the Calakmul hegemony at this time, and was probably dominated by Piedras Negras.

In the same tenor the hieroglyphic inscriptions point towards a political alliance since early times between Yaxchilan and Bonampak. Yaxchilan Lintels 49, 37, and 35 (Figure 5-2b, 5-2c, and 5-2d,) record the visit of dignitaries from Bonampak at ca. AD 416, 504, and 526 respectively. Yaxchilan however seems to always have played the dominant role. Sculpture Stone 4 of Bonampak substantiates this assumption. In this monument Itzam-Balam (I?) Ch'ul Ahaw from Yaxchilan is depicted overseeing the accession of his Bonampak counterpart Chan Muwan I, at around 9.8.6.13.7 (AD 600) (Figure 5-5). Interestingly, Chan Muwan included amongst his titles the emblem glyph of Lacanha, opening the possibility that these two polities formed a joint kingdom during this period.

Figure 5-5 Itzam-Balam of Yaxchilan oversees the accession of Chan Muwan I of Bonampak (Bon. S.S.4, from Schele and Grube 1995)

Three years after his accession on 9.8.9.15.11 (AD 603) Chan Muwan makes war on Palenque, and on 9.8.17.15.0 (AD 611), Calakmul assisted by Pomona wages war on Palenque once again. Finally, Piedras Negras begins its hostilities against Palenque in 9.9.11.12.3 (AD 624), when Ruler 1 captures Ch'ok Balam a Yakun ahaw (subservient lord) of the Palenque king and the Sak Tz'i' king K'ab' Chan Te'' (PN St.26, Figure 5-6). These events show that in a period of 25 years Calakmul and its allies were putting the pressure on Palenque keeping it at bay perhaps to prevent it from assisting Tikal, which had just began to emerge from its long hiatus (9.5.0.0.0-9.8.0.0.0; AD 534-593).

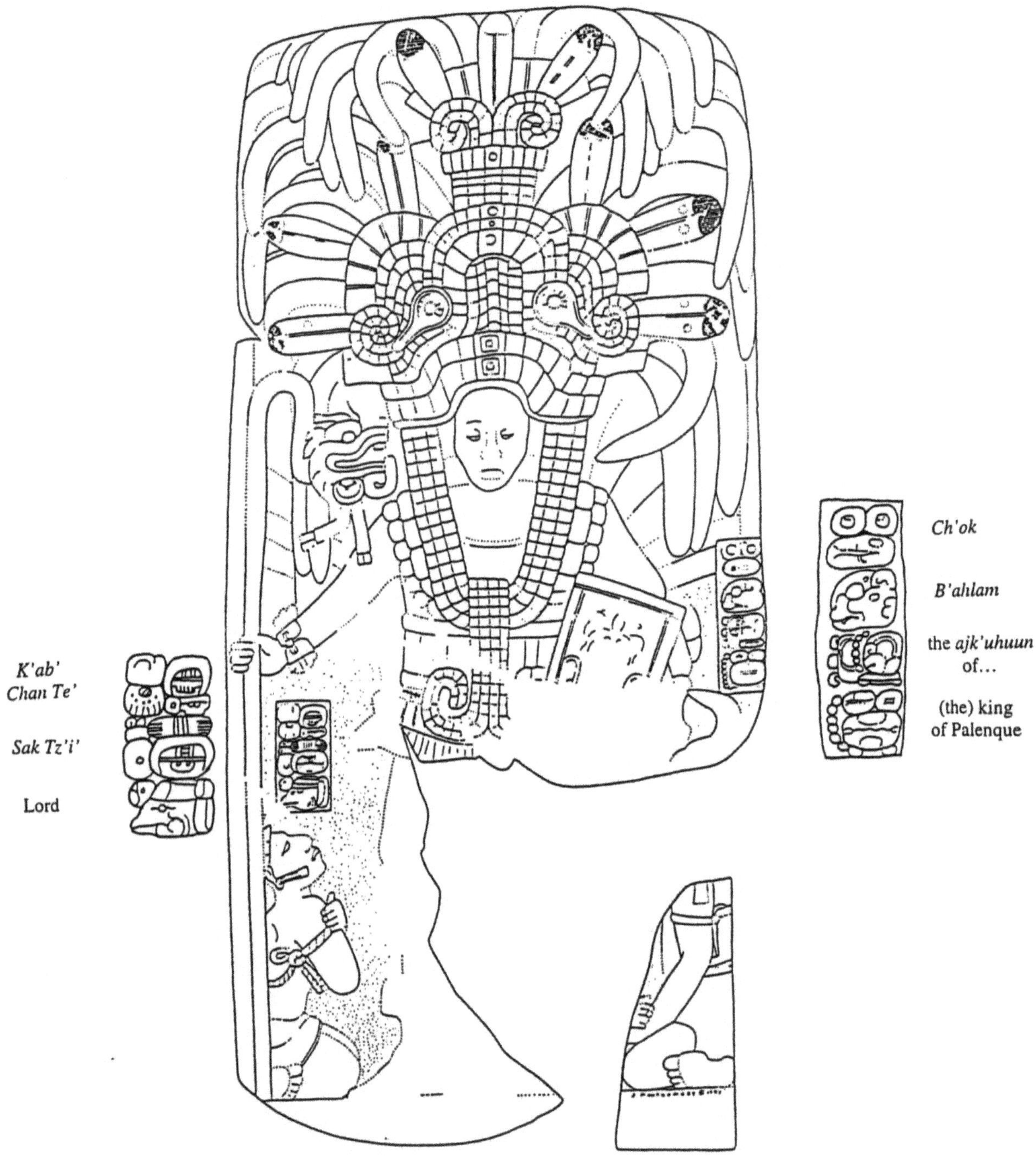

Figure 5-6 Ruler 1 of Piedras Negras stands over his Palenque and Sak Tz'i' captives (PN St. 26 drawing by John Montgomery)

Sak Tz'i' is mentioned for the first time in Piedras Negras Stela 26, in connection with the Palenque lord capture. This simultaneous capture implies that Sak Tz'i' may have been a close ally in the region of Palenque, and by extension of Tikal. The events that follow support this idea. K'ab' Chan Te'' seems to have been released by Ruler 1 of Piedras Negras probably on the condition that he falls in line and join the Calakmul alliance. However as we shall see K'ab' Chan Te'' had a mind of his own. Figure 5-7 shows the proposed political map of the region during this period.

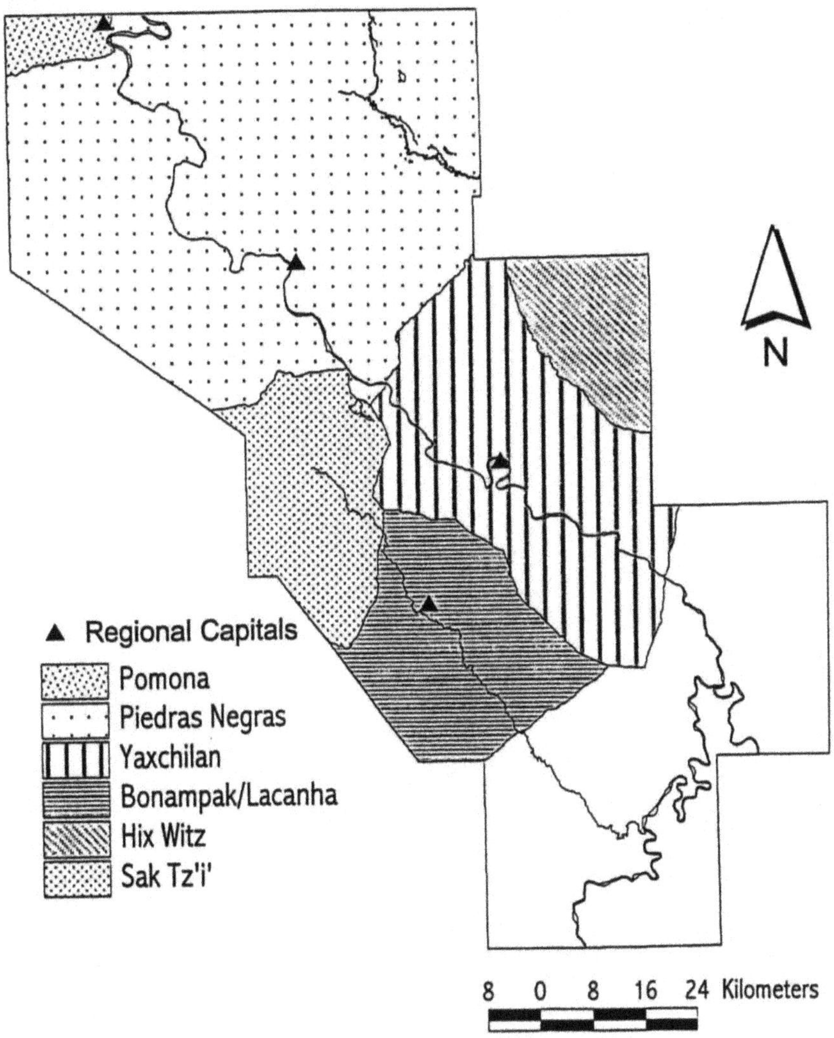

Figure 5-7 Suggested territorial extents of the Upper Usumacinta kingdoms at AD 599-624

The Sak Tz'i' Years: 9.10.8.3.5 to 9.11.16.11.6 (AD 641-669)

As mentioned above it would seem as though the Sak Tz'i' king was captured only to be released later by Ruler 1 of Piedras Negras as it becomes apparent from his mention in two inscribed monuments at a later date, the Denver and Brussels Panels (Figure 5-8). Although his release must have been conditioned, in 9.10.8.3.5 (March 14, AD 641) K'ab' Chan Te'' does something to Ruler 1 that starts an outburst of war frenzy. The nature of K'ab' Chan Te''s actions is not very clear but it must have been of serious proportions since as a result Ruler 1 of Piedras Negras sends his La Mar vassal Nik-Mo' after him. Nik-Mo' reaches the Sak Tz'i' capital putting it to the torch but K'ab' Chan Te'' manages to escape. The account of the events to follow are a clear indication that the kingdom of Bonampak, a close neighbour of Sak Tz'i', participated in the military campaign against the latter, thus making clear that Bonampak's loyalties were with Piedras Negras and the Calakmul alliance.

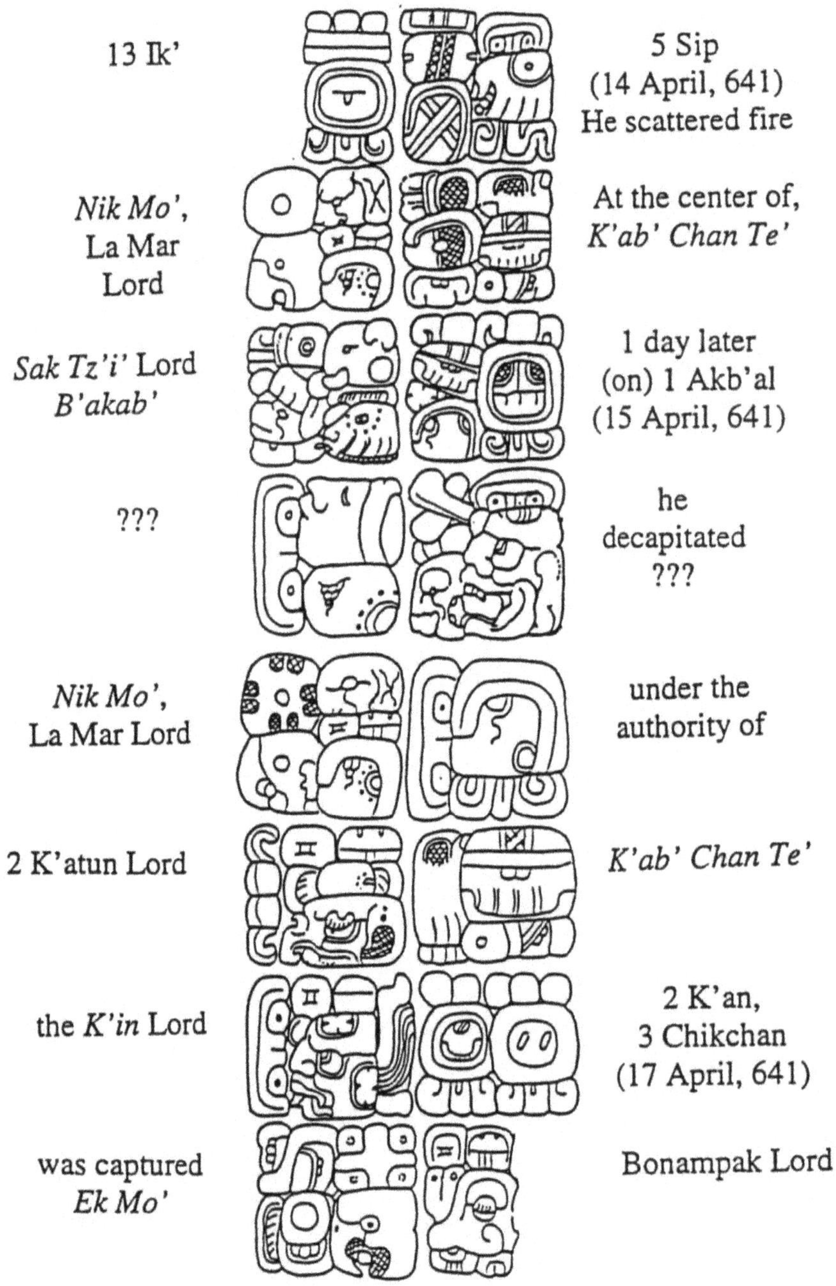

Figure 5-8 Account of the Sak Tz'i' wars on the Denver and Brussels Panels (Adapted from drawing by Linda Schele)

A day after the attack on his city (March 15, 641) K'ab' Chan Te' retaliates attacking the centre of La Mar where he "axes" something or someone (Schele and Grube 1994:116). Two days later (March 17, 641) K'ab' Chan Te' captures Nik Mo' and Ek Mo', a Bonampak ahaw. All available evidence indicates that Sak Tz'i' prevailed in its war against the Piedras Negras-Calakmul alliance providing in the process a foothold in the region for Palenque and Tikal. It is very likely that Sak Tz'is victory would have curtailed access to Piedras Negras, Yaxchilan, Bonampak and Lacanha to at least part of the routes of communication located on the western side of the Usumacinta River. If we look back at Figure 4-11, which shows the proposed territorial extent of Sak Tz'i', we'll see that this area extended towards the Usumacinta at the strategic location of El Chile and El Cayo, and a short distance across from the river.

Yaxchilan's king, Yaxun Balam III captures on 9.10.14.13.0 (AD 647) a Hix Witz ahaw (Yax. H.S.3). This kingdom was located towards the east, on the road to the Peten so it is possible that Yaxun Balam was trying to secure his kingdom's connection with the Peten. Sak Tz'is control over the western routes continued until at least 9.11.16.11.16 (AD 669); when apparently Piedras Negras attacked El Cayo then under the aegis of the Sak Tz'i' (PN St. 37).

Before that date however Sak Tz'is supremacy is suggested by the following accounts: Lintel 2 of Piedras Negras dated on 9.11.6.2.1 (AD 658) depicts six young ahaws coming from Lacanha, Bonampak and Yaxchilan to pay homage to Ruler 2 and his young heir (Figure 5-9).

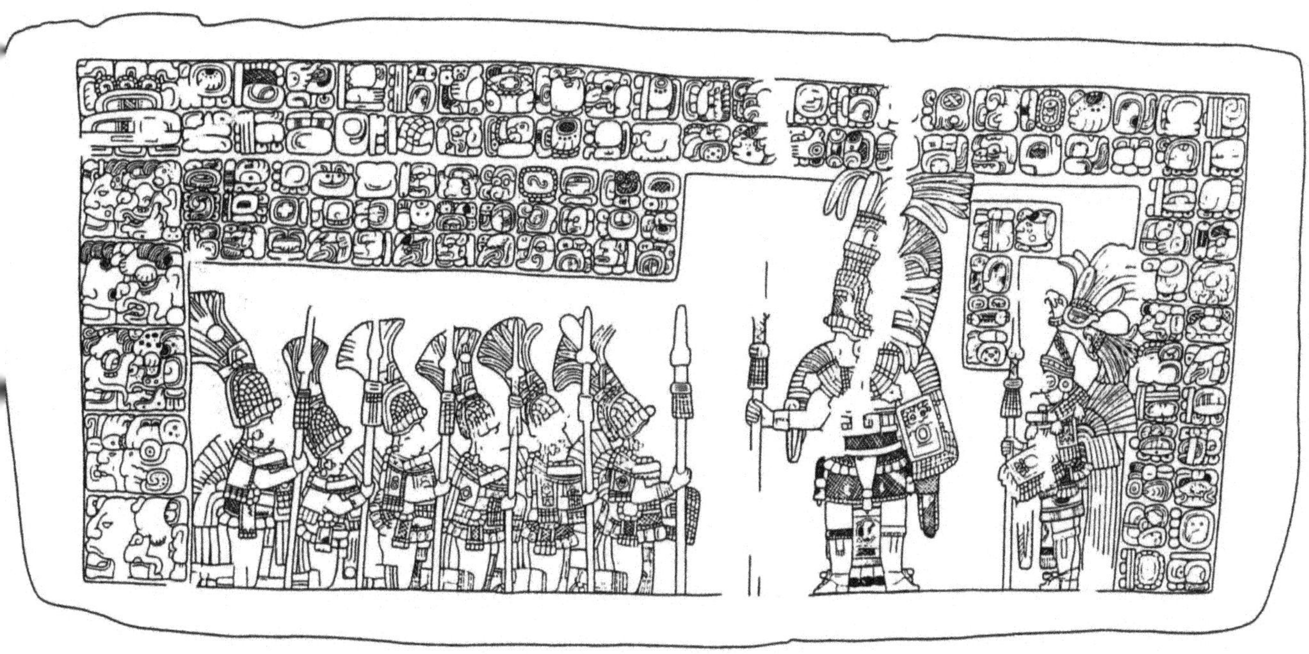

Figure 5-9 Lacanha, Bonampak, and Yaxchilan lords paying homage to Piedras Negras Ruler 2 and his heir (PN L. 2 drawing by David Stuart)

This scene, which undoubtedly represents a statement of political alliance is also revealing in another sense. If Sak Tz'i' was limiting passage on the western side of the Usumacinta (see Figure 5-1), then Bonampak and Lacanha would be forced to take alternative routes to Piedras Negras. In this context it is likely that the young princes took the route to Yaxchilan and from there descended either downstream to Piedras Negras risking the rapids, or used the alternative inland route. Another possibility is that these young ahaws were "guests" of Piedras Negras in a very similar fashion as the Cocom family of Post-Classic Mayapan hosted in their city the heads of lineage of the neighbouring polities in order to ensure their loyalty. Whatever the case it seems that Sak Tz'i' had in fact the upper hand in this arrangement.

It is also during this period of time that the Tikal king Nu-Bak-Chak passes through the region. Nu-Bak-Chak was fleeing from Dos Pilas after the allied forces of Dos Pilas and Calakmul had retaken this city. On route to Palenque in 9.11.6.16.11(AD 659), Nu-Bak-Chak and his fleeing army had a skirmish with Yaxchilan, from which they emerge victorious, managing to capture Itzam Balam's war palanquin (Pal. House C). For Nu-Bak-Chak to overcome the Yaxchilan forces in their own territory with an army that was presumably still healing from its defeat by Calakmul and Dos Pilas, it would have represented a task of epic proportions. It is therefore unlikely that Nu-Bak-Chak may have accomplished this feat alone. Nu-Bak-Chak must have had a powerful ally to assist him in his exodus. The obvious candidate is Sak Tz'i', an old time ally of Palenque, hence part of the Tikal alliance. Figure 5-10 shows the possible route that Nu-Bak-Chak and his forces may have taken. From this figure we can see that for the most part, this route passes through territory controlled by Sak Tz'i'.

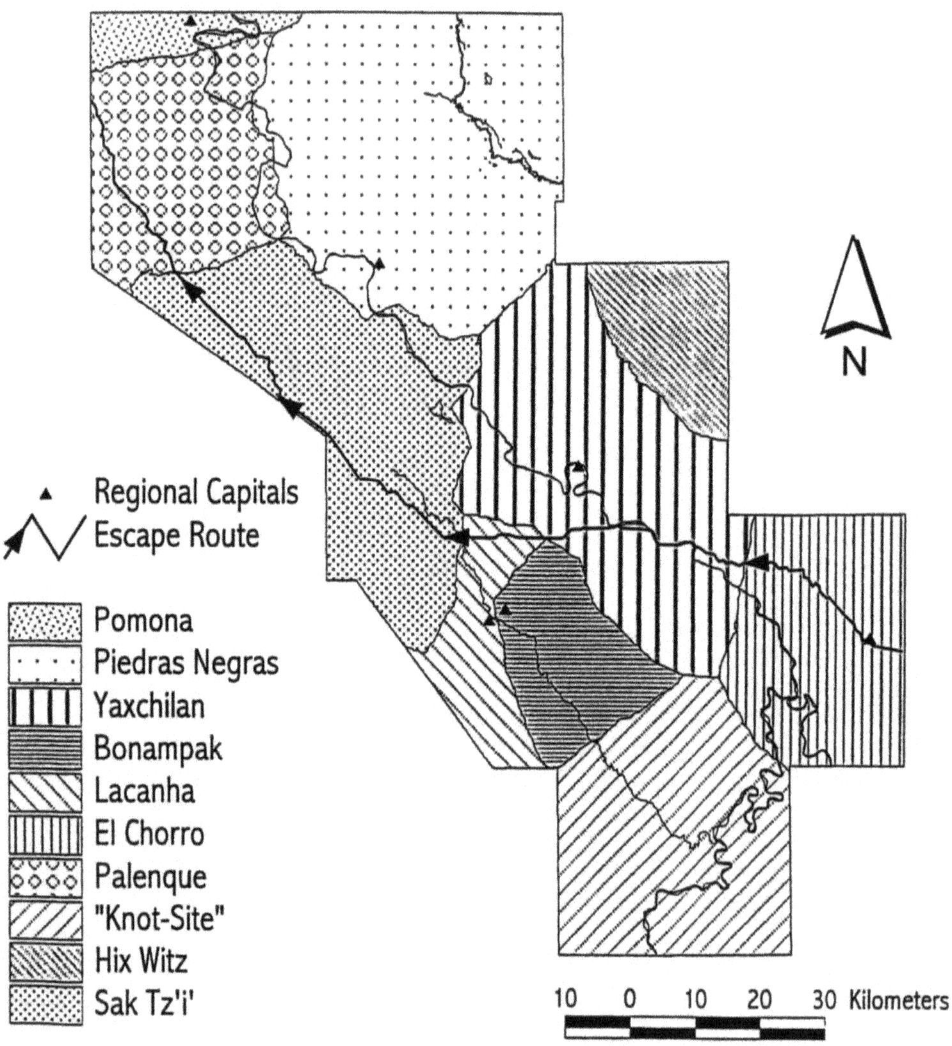

Figure 5-10 Proposed escape route of Tikal's king Nu-Bak-Chak passing through Sak Tz'i's territory at AD 659

Another possible Tikal ally in the region was either the "Knot-Site" polity, or El Chorro. Descending from the Pasion region Nu-Bak-Chak must have crossed the territories controlled by the latter polities. There is circumstantial epigraphic evidence that leans the balance in favour of the "Knot-Site" polity. First, almost three years after Nu-Bak-Chak's passage, in 9.11.9.15.1 (AD 662) Balah Kan K'awil the Dos Pilas king wages war on the "Knot-Site" (Dos Pilas, Hieroglyphic Stairway 2). This war could have been in retaliation for this polity's support to the fleeing Tikal king. In any case what this attack indicates is that the "Knot-Site" polity was not in good terms with the Calakmul alliance. Second, the "Knot-Site" would become a very important ally to Sak Tz'i' towards the second half of the 14th katun when both polities control the Bonampak/Lacanha territory.

During these events Piedras Negras was probably too preoccupied concentrating its forces to defend its northern flank from Palenque. Apparently, in co-ordination with Nu-Bak-Chak's passage through the region, Palenque attacked Pomona capturing one of its prominent nobles (Palenque, House C, Hieroglyphic Stairway). Figure 5-11 shows the proposed territorial distribution for this period.

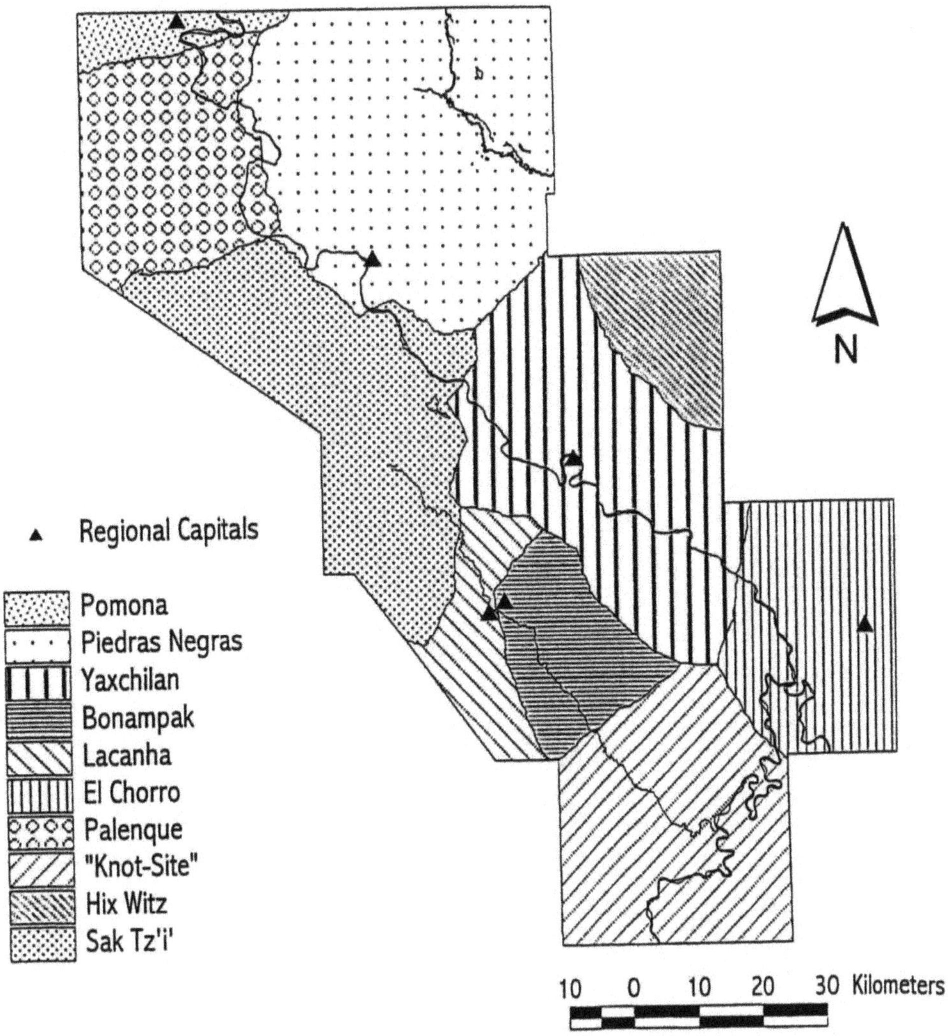

Figure 5-11 Suggested territorial extents of the Upper Usumacinta kingdoms at AD 641-669

The Calakmul Alliance Wavers: 9.11.16.11.6-9.14.10.5.0 (AD 669-722)

Piedras Negras' woes seem to come to an end with their attack on El Cayo in 9.11.10.16.17 (AD 669) after this Sak Tz'i' may have withdrawn from El Cayo, although by no means definitively vanquished. The first part of this period was more or less a time of healing and political readjustments. As in the past, the inscriptions of this sub-period deal with war related events but they also give prominence to marriage and political alliances.

In Yaxchilan Itzam Balam "The Great" is ruling. This king started a campaign that would lead his kingdom to complete independence from Piedras Negras. Not that Piedras Negras ever wielded effective control over Yaxchilan, but rather Yaxchilan's dealings with Calakmul seem to have been mediated by the former in the past. In 9.12.8.14.1 Itzam Balam captures his most famous prisoner Ah Nik from Man (Figure 5-12). This act would become a matter of contention for Piedras Negras who apparently had vested interests in this site (i.e. the influential Lady Kaltun, principal wife of Ruler 3, came from this site). Perhaps in an effort to dampen grievances between his two main regional allies the Calakmul king sends his envoys to Piedras Negras. Ruler 2 received this group of royal emissaries from Calakmul in 9.12.13.4.3 (AD685). This visit may be interpreted as an implicit recognition of Piedras Negras regional seniority.

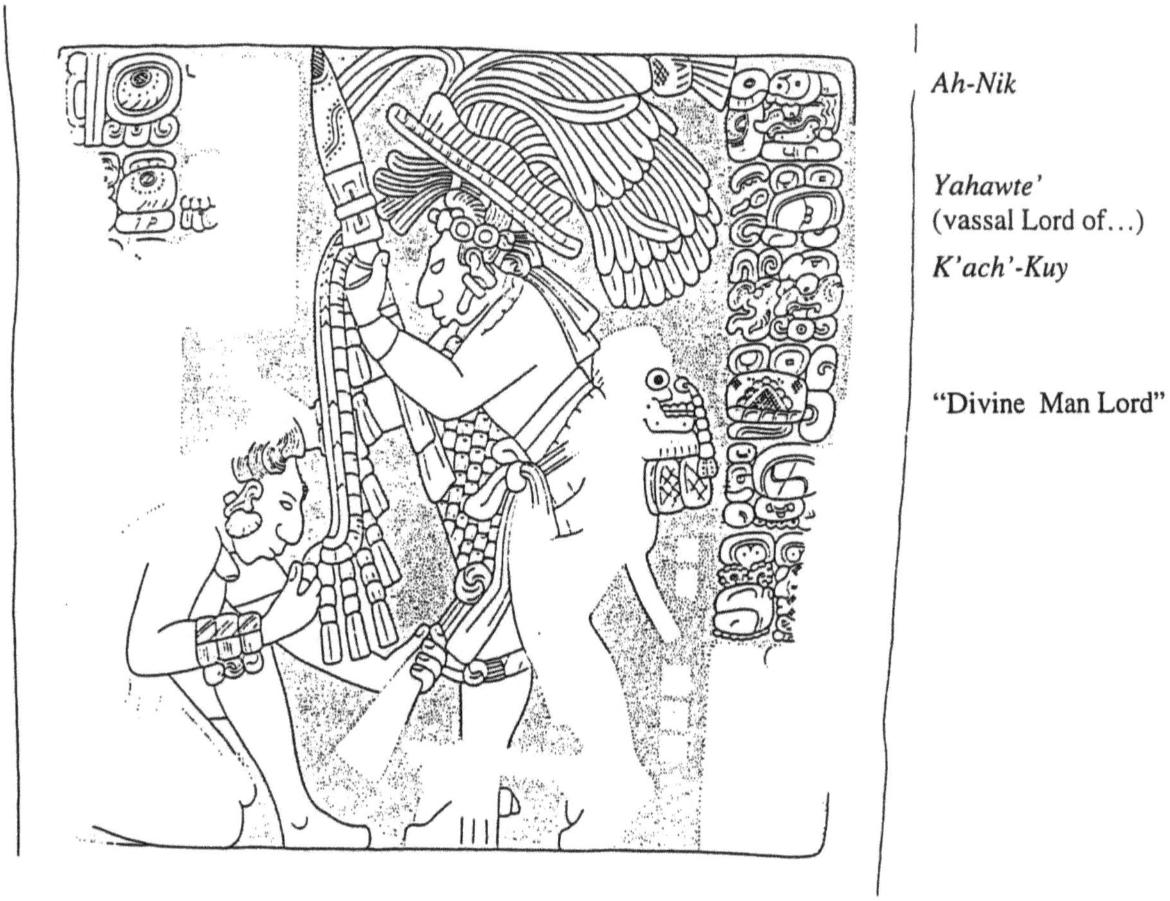

Figure 5-12 The capture of Ah-Nik from Man by Itzam Balam of Yaxchilan (Yax. L. 45 drawing by Ian Graham)

By the end of the 13th katun a third party in this hegemonic power struggle makes its entry into the region: Tonina. In 9.13.19.13.3 (AD 711) Bak-Nal-Chak the Tonina king initiated his military campaign outside his own sphere by capturing Palenque's king K'an-Hok Chitam (Figure 5-13a). Four years after this capture Etz'nab-Ch'oy a Bonampak ahaw and probably its ruler, declares himself vassal (yahaw) of Bak-Nal-Chak of Tonina (Figure 5-13b).

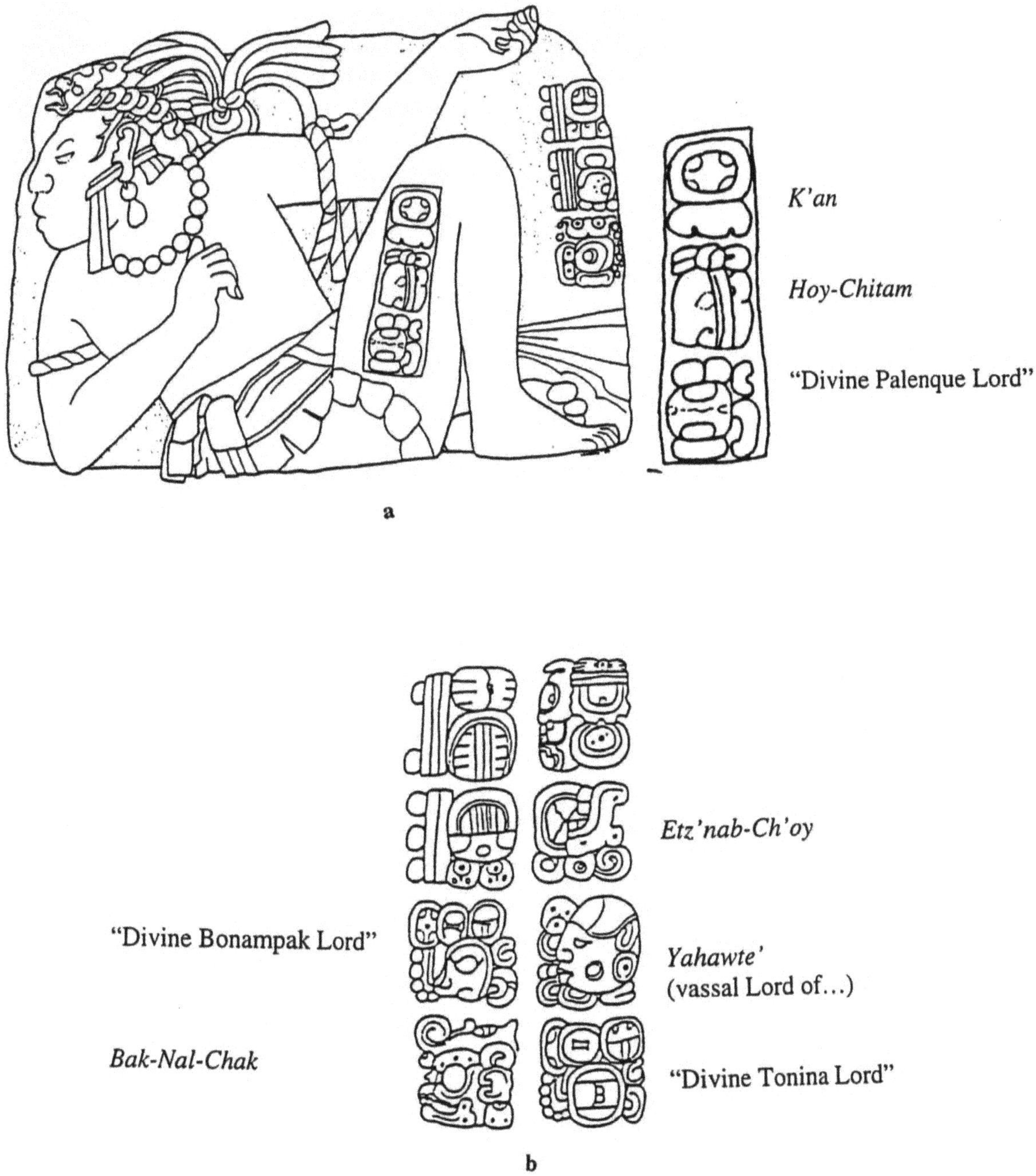

Figure 5-13 Tonina's intervention in the Upper Usumacinta: (a) K'an-Hok Chitam king of Palenque displayed as as a captive in Tonina (Ton. Mon. 122 drawing by Ian Graham); (b) Etz'nab-Ch'oy from Bonampak declares himself vassal of Bak-Nal-Chak (St. Louis Hieroglyphic Column drawing by Linda Schele)

The intervention of this third party obviously disrupted the power balance of the region. Probably due to its geographic isolation, Tonina does not seem to belong to either of the hegemonies. Tonina's incursion, although very disruptive, was brief. Logistics, in terms of distance from their centre may have played the determinant role. It is reasonable to assume that, in order to penetrate the region more effectively, Tonina may have sought an alliance with a local force. Sak Tz'i' is the most likely candidate, in turn Sak Tz'i' probably saw in this alliance the opportunity to regain its old dominion after being pushed back by Piedras Negras.

In relation to Tonina's brief presence, it is tempting to speculate that after aiding its foreign allies Sak Tz'i' withdrew its support, taking over Tonina's conquered lands.

No inscription attesting this alliance has been recovered, however it is interesting to note that on a later date, 9.18.0.0.0 (AD 790) Tonina displays a Sak Tz'i' prisoner, perhaps in retaliation for the past defection (Figure 5-14). There is the possibility that Sak Tz'i' may have returned to the Tikal hegemon. In the extra-regional scene at around this time, Tikal, with the accession to the throne of the audacious king Hasaw-Chan K'awil, had regained its prominence, and had turned the table on Calakmul. Hasaw-Chan K'awil inflicted a decisive defeat on his Calakmul counterpart in revenge for the latter's involvement in his father's death Nu-Bak-Chak, occurred at the hands of Balah-Kan-K'awil, king of Dos Pilas. As a consequence the Calakmul alliance suffers a considerable weakening.

Figure 5-14 A Sak Tz'i' captive displayed at Tonina (Ton. Mon. 83 drawings by Peter Mathews [a], and Nikolai Grube [b])

After Tonina is gone Sak Tz'i' seems to have established an alliance with the "Knot-Site" polity "turning the heat" on Yaxchilan and Piedras Negras, the two most powerful Calakmul allies in the region. This idea seems to be substantiated by the available epigraphic evidence, Panels 1 and 2 from Nuevo Jalisco record the seating in 9.14.10.5.0 (AD 722) of a Lacanha ruler, Tab Balam of the "Knot-Site" under the aegis (*u-kabhi*) of the Sak Tz'i' king. Likewise the Zurich Panel refers to a Lacanha Lord, K'inich Chak Chi as a vassal of K'ab-Chante of Sak Tz'i' (Figure 5-15a). Finally, the New York City Panel records a house dedication event performed by K'ab-Chante, Ch'ul Bonampak and Sak Tz'i' ahaw (Figure 5-15b). These monuments clearly indicate that both Bonampak and Lacanha have been taken over by Sak Tz'i' and the "Knot-Site" polity respectively.

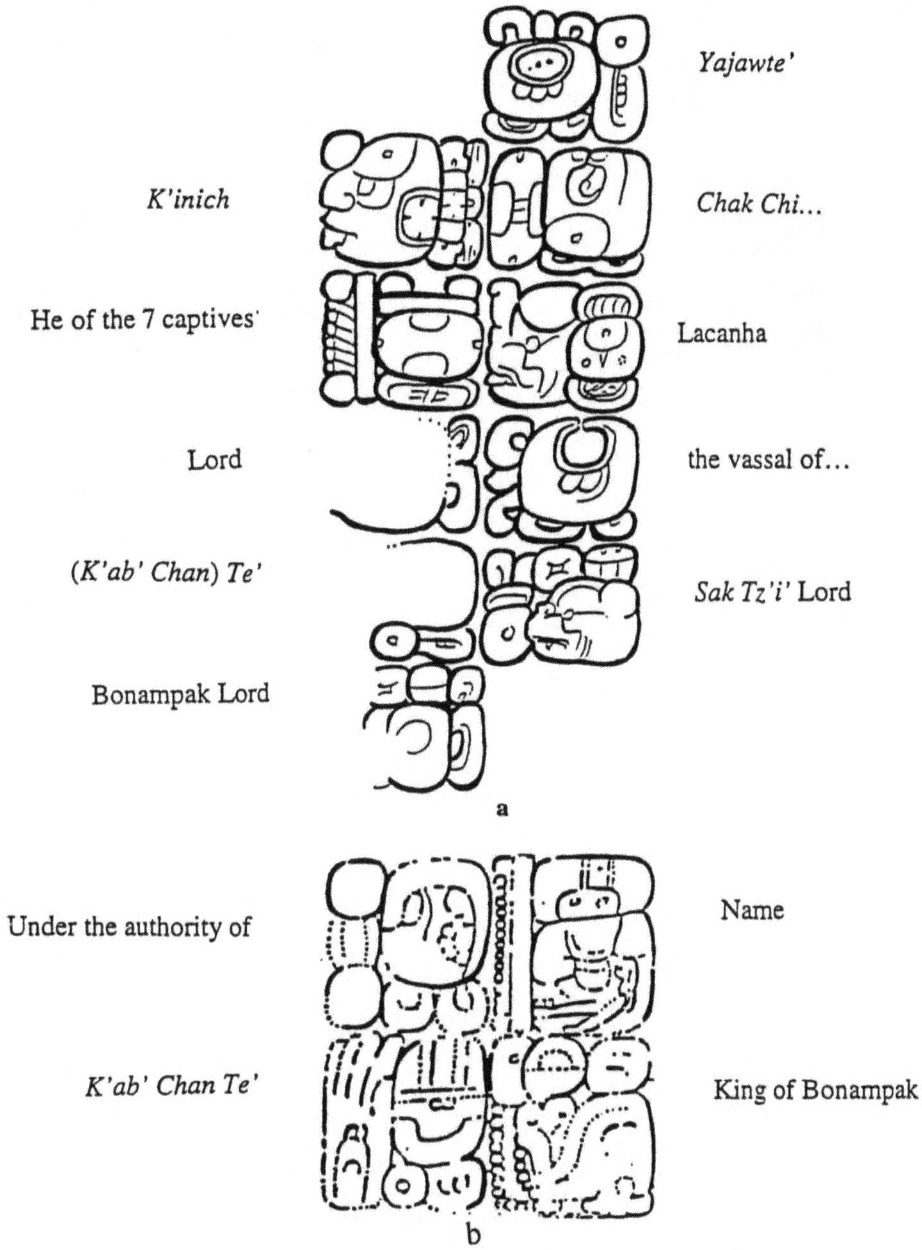

Figure 5-15 Sak Tz'i's control over Lacanha and Bonampak (a) Zurich Panel; (b) New York City Panel (adapted from drawings by Ian Graham [a] and John Montgomery [b])

Yaxchilan was too preoccupied with problems of its own to aid its old ally. Perhaps the capture of Ah K'an U-Sih, Buk-Tun ahaw ocurred in 9.14.1.17.14 (AD 713) (Yax. Lint.46) marked the beginning of Yaxchilan's woes. Although the whereabouts of Buk-Tun are unknown Mathews has commented on the possibility that it was located in the vicinity of Lacanha since the reference to this site is framed in terms of earlier captures from Lacanha (Mathews 1988:163). By 9.14.11.4.14 (AD 723) things were getting pretty rough for Yaxchilan. Not only were Sak Tz'i' and the "Knot-Site" pushing on its western flank, Itzam K'awil of Dos Pilas had captured one of Itzam Balam's ahaws (D.P. HS 1-III). Thus possibly this capture was marking an attempt by Dos Pilas to encroach on Yaxchilan's southern frontier. It is perhaps due to these pressures that Itzam Balam has a change of heart and considers breaking away from the Piedras Negras/Calakmul alliance to join Tikal.

Sometime between 9.14.0.0.0 and 9.14.13.0.0 Itzam Balam pays a visit to the Tikal king Hasaw-Chan-K'awil. Guenter (1999) noted that one of the individuals depicted in cylinder 4P-8/2 of Tikal carries the title *u'chan Ah-Nik*, "the guardian of Ah-Nik", which is none other than Itzam Balam using his favourite captor title (Guenter and Zender 1999, unpublished manuscript). In their paper Guenter and Zender suggest that perhaps as an outcome of this meeting Itzam Balam rebels against Piedras Negras waging an unsuccessful war against the latter. This change of alliance however must have "ruffled a few feathers" amongst those nobles that had a vested interest in Calakmul, and so intrigues must have started to brew.

Surely enough when Itzam Balam's mortality started to become more apparent the speculations and intrigue in relation to his successor must have intensified. A series of sculptured monuments from Yaxchilan suggest that little less than a year before Itzam Balam's death, Yaxun-Balam IV the king-to-be, concerts a pact with Calakmul. Lintels 39, 35, and 14 record a vision serpent rite --which was usually a pre-accession ritual-- with the participation of Yaxun Balam, his influential wife and brother-in-law Lady and Lord Chak-Chami and his Calakmul wife Lady Ik-Kimi. The whole event takes place in Yaxchilan under the auspices (*u-kabhi*) of Yukom of Calakmul. It may either be that the old king never forgave Yaxun Balam for backing down on his word, and decided to designate a different successor, or that this designation obeyed pressures from the competing noble factions.

After Itzam Balam's death, which occurred sometime after 9.15.10.17.14 (AD 742), there is a rather obscure ten-year interregnum period that takes place at Yaxchilan. Yaxun Balam did not accede to the throne until 9.16.1.0.0 (AD 752). More recent re-interpretations of Piedras Negras Lintel 3 (Schele and Grube 1995), suggest that there might in fact have been another ruler in Yaxchilan prior to Yaxun Balam IV. This monument records a Balam ahaw from Yaxchilan paying an official visit to the Piedras Negras king precisely during the Yaxchilan interregnum period, I will address this important event in the following section.

By the end of this period, Yaxchilan remained within the Calakmul hegemon, however an important change took place. Now Yaxchilan was dealing directly with Calakmul instead of doing it through Piedras Negras' mediation. This change becomes more apparent in the events that took place during the next period. Figure 5-16 shows the proposed territorial extent of these polities during this period.

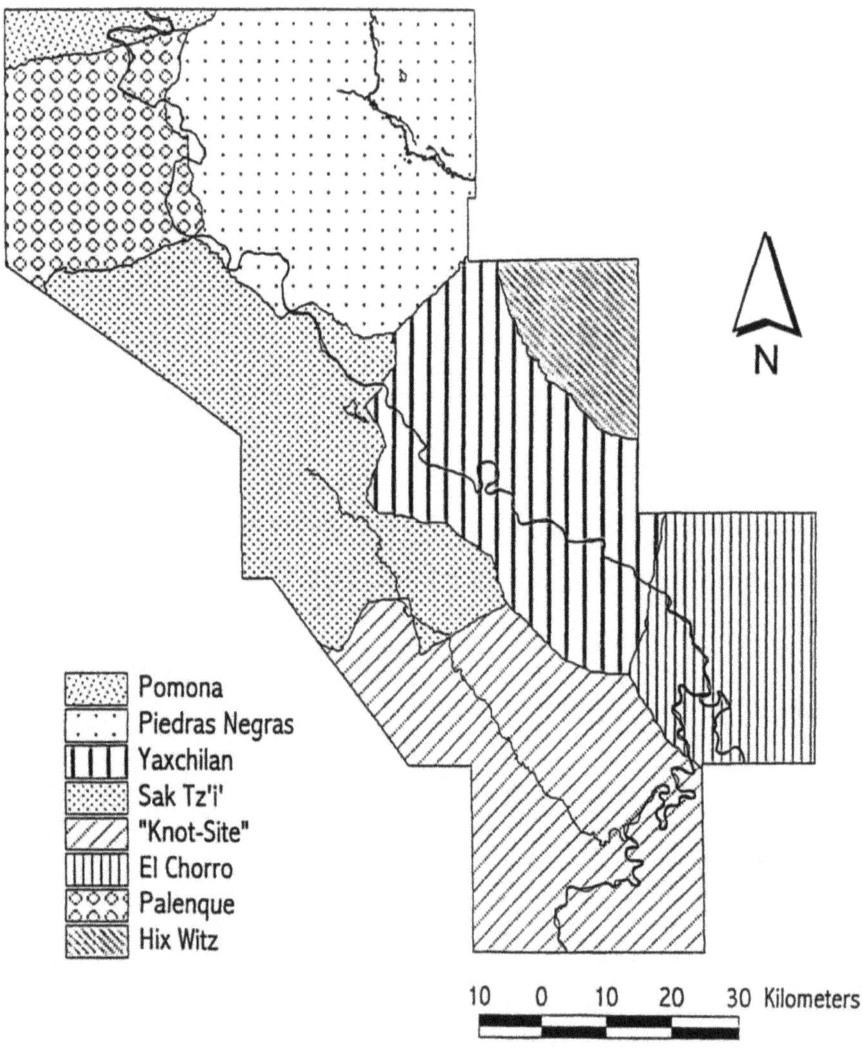

Figure 5-16 Suggested territorial extents for the Upper Usumacinta kingdoms at AD 669-722

The Yaxchilan Expansion: 9.15.19.2.2 to 9.16.8.3.18 (AD 750-759)

The inscriptions corresponding to this period suggest an expansion of Yaxchilan with a concomitant contraction of Piedras Negras influence. While Yaxun Balam IV rallies the support of Yaxchilan's most important sahals and allies on his eastern front, Piedras Negras concentrates its efforts in guarding its northern flank from Palenque. The text contained in a new stela from Pomona attests that this site was still controlled by Palenque as K'inich-Ho-Ix, Pomona king celebrated the 9.16.0.0.0 period ending in the company of K'inich Kan-Balam royal envoy from Palenque. There are also indications that Sak Tz'i' is once again in control of El Cayo (El Cayo Panel 1, see below).

At Yaxchilan with Calakmul's blessings, the power balance begins to tip in favour of Yaxun Balam who has already assumed some royal functions (see for example Bon. S.S.5, Yax.Alt. 9, HS 4-III in Appendix 3). I have already mentioned that Schele and Grube (1995) noted that Piedras Negras Lintel 3 records a royal visit that took place in 9.15.18.3.13 (AD 749) paid to Ruler 4 by Sak-Hukub Yat-Balam Ch'ul Yaxchilan ahaw. This visit is framed in a similar manner as other royal visits, and it took place during the Yaxchilan interregnum. It is then reasonable to think that Sak-Hukub Yat-Balam may be the actual Yaxchilan king whose main purpose at Piedras Negras was to secure the support of Ruler 4. Sak-Hukub Yat-Balam was probably increasingly apprehensive of Yaxun Balam's efforts to rally supporters and wanted to secure for himself a strong ally in the event of a "coup" and/or a place where he could eventually seek refuge. However, Sak-Hukub Yat-Balam was never mentioned in any of the known Yaxchilan inscribed monuments. Nevertheless, as Schele and Grube rightly point out, the reason why there are no references to this individual in Yaxchilan was because Yaxun Balam shaped the entire history of this period (Schele and Grube 1995:111).

Yaxun Balam did not accede into power until 9.16.1.0.0 (AD 752). Once in office he continued reinforcing his ties with his eastern allies as the participation of his wives from Hix Witz and Motul de San José suggest (Yaxchilan Lintels 13, 17, 5, 42, 6, 43, and15). In this way Yaxun Balam was guaranteeing Yaxchilan's access to the Peten road and by extent direct access to Calakmul. On a related issue, the war preparation rituals in which he participated along with his eastern sahals from La Pasadita and perhaps Site R suggest --although there is no conclusive evidence-- that he may have decided to harass the Piedras Negras king by raiding a series of settlements within his territory.

On 9.16.6.10.1 (AD 757) Yaxun Balam finally decides to pay a visit to the Piedras Negras king (PN L.3). The details of the matters discussed during that summit are unclear. It is possible however that what Yaxun Balam was seeking was the official recognition by Piedras Negras of his right to rule (and perhaps the surrender of the deposed Yaxchilan king), while Ruler 4 probably sought a "non-aggression pact". The text also gives the impression that the old king was "reprimanding" the arrogant young king. The text is a relation of the mutual interests that have linked both polities through time. Ruler 4 reminds Yaxun Balam how through the doings of his predecessors his grandfather Yaxun Balam III was able to rule at Yaxchilan (Schele 1996 personal communication).

Twenty days after this meeting took place, in 9.16.6.11.0 Yaxun Balam appears in Yaxchilan with four prisoners being presented to him by a subordinate lord (perhaps the deposed king and his entourage) (Figure 5-17). Although most of the text is eroded the date of the monument implies, as Schele and Grube suggest, that the Piedras Negras summit had war implications. Reinforcing this idea is Ruler 4's death 18 days after Yaxun Balam's captives were presented. Schele and Grube suggest that his death might have been related to wounds received in battle (Schele and Grube 1995:122).

Whether Ruler 4 died as the result from wounds received in a battle fighting against, or side by side with Yaxchilan, or whether he died of natural causes (he had just turned 56) is still an unresolved matter. However the fact that Yaxun Balam's kingship was officially recognised by Piedras Negras five days after Ruler 4's death suggests to me that Piedras Negras and Yaxchilan had reached an agreement. The official recognition however, was phrased with an air of contempt towards the Yaxchilan king. Lintel 3 of Piedras Negras records as the official date of accession of the Yaxchilan king the date 9.16.6.12.2 (December 1, AD 757). On this day Yaxun Balam of Yaxchilan (without the "holy ahaw" title) acceded into ahawship "under the auspices of" Ruler 4 of Piedras Negras (Schele and Grube 1995:122).

Yaxun Balam spent the rest of his rule consolidating his kingdom's eastern frontier reinforcing his alliances with his sahals of La Pasadita, Site-R and Laxtunich as well as with the kingdoms of Hix Witz and Motul de San Jose. It would be up to his son Chel-Te-Chan (Itzam Balam III) to win back the western frontier from Sak Tz'i'. Figure 5-18 shows the reconstruction of the polity territorial distribution for the period described above.

Figure 5-17 Yaxun Balam is presented with four captives (Yax. L. 12 drawing by Ian Graham)

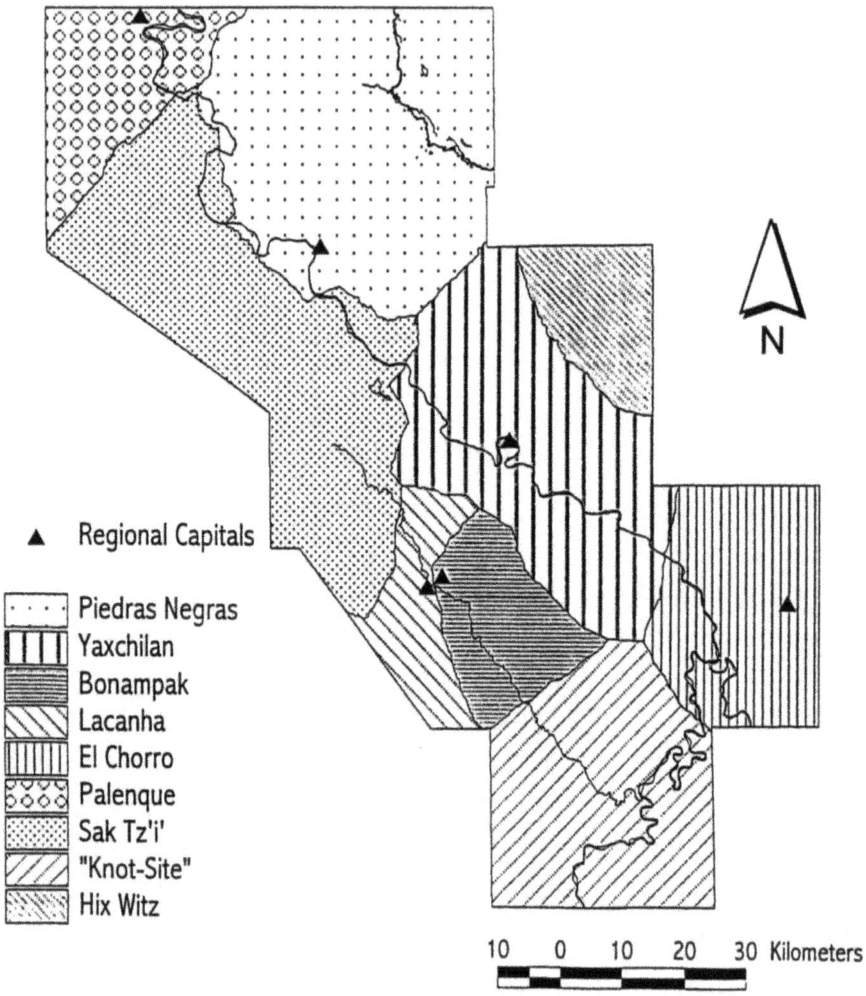

Figure 5-18 Suggested territorial extents at AD 750-759

The Reclaiming of the Western Frontier: 9.16.12.2.6 to 9.17.1.5.9 (AD 763-772)

This period although well documented is a bit confusing, mainly due to the shifting regional alliances. During this period old foes become allies, old allies become foes and all of them write their own account of history. At El Cayo the local sahal acknowledges in 9.16.12.2.6 (AD 763) as his overlord the Sak Tz'i' ahaw (El Cayo Pan.1), who in turn recognizes Ruler 5 of Piedras Negras as his overlord. This expression of submission is repeated again in 9.17.1.5.9 (AD 772) by another El Cayo sahal, Chan-Panak Wayib. The text is a bit puzzling since as we recall in the last record we had on Sak Tz'i' (the New York Panel dated ca. AD 726), this polity was calling the shots along most of the western bank of the Usumacinta River.

What were the changes that occurred that compelled Sak Tz'i' to recognise a subservient position to his long time enemy Piedras Negras? Interestingly after 9.15.15.0.0 (AD 746) we do not find any more references to the "Knot-Site" polity. It is possible then, that this polity was forced out of the scene by affairs brewing outside the region. Dos Pilas Hieroglyphic Stairway 1, step III records 9.15.16.6.9 (AD 747), as the date when the Dos Pilas king K'awil Chan-K'inich captured Chak-Chan-ha-Xoc of Yaxchilan (this capture took place during the confusing years of the interregnum). It could be that this Dos Pilas king took advantage of the chaos that was prevailing at Yaxchilan and made an attempt to expand his dominion towards this direction distracting in the process the "Knot-Site" polity from its alliance obligations.

The events that took place in the previous period strongly suggests that after Yaxun Balam's accession Yaxchilan grew stronger while at the same time Piedras Negras lost ground. In this train of events it is possible that both Piedras Negras and Sak Tz'i' realised that a strong Yaxchilan represented a threat to their own survival and saw in their alliance a strategic move. The fact that the El Cayo sahals recognise that their authority comes from the Sak Tz'i' lords (*u-kabhi*) and not from their Piedras Negras overlord, is indicative that the latter did not overcome the former in battle, but that their alliance came out of mutual agreement. Furthermore these expressions also suggest that although Sak Tz'i' accepts the authority of Ruler 5, El Cayo is still under the control of Sak Tz'i'. The events to follow give credence to this scenario.

Yaxun Balam dies sometime after 9.16.17.6.12 (AD 768), date of his last known official act recorded in Yaxchilan Lintel 9. His successor Chel-Te Chan K'inich assumes power shortly after 9.16.18.0.19 (AD 769). Once in office, Chel-Te concludes his father's efforts to consolidate his kingdom's domain and sets forth in a campaign to recover the western frontier. As mentioned earlier, Yaxchilan and Bonampak had a long history of political alliance dating to at least as early as AD 416. Thus expectedly Chel-Te would provide assistance to its old ally in order to rid them of the Sak Tz'i' yoke. Jointly these two polities mounted a successful military campaign against the latter. In 9.17.16.3.8 (January 4 787) Chel-Te of Yaxchilan captures a Sak Tz'i' lord, and four days later his ally Chan Muwan II follows suit capturing another Sak Tz'i' ahaw (Figure 5-19a-b). With these captures Yaxchilan and Bonampak were not only regaining their occupied territories they were also turning once more the heat on Piedras Negras by attacking one of its newly gained allies.

Piedras Negras however, was too involved in its own projects of regaining its old lustre to be preoccupied by Sak T'si's woes. Besides, Sak Tz'i' had always been a thorn in the foot of Piedras Negras, and the latter saw perhaps in this defeat the opportunity to grab back its lost territories. The same year that Yaxchilan and Bonampak were attacking Sak Tz'i', Piedras Negras was probably involved in probing Palenque's strength now that it did not have its regional ally. In 9.17.16.14.19 (August 23, AD 787) Anabil Ah ?K'in K'ul Tok was captured by Piedras Negras. Schele and Grube observed that this individual was the vassal lord of another lord that used amongst his titles a toponym that is similar to a toponym found at Palenque (Schele and Grube 1995:155).

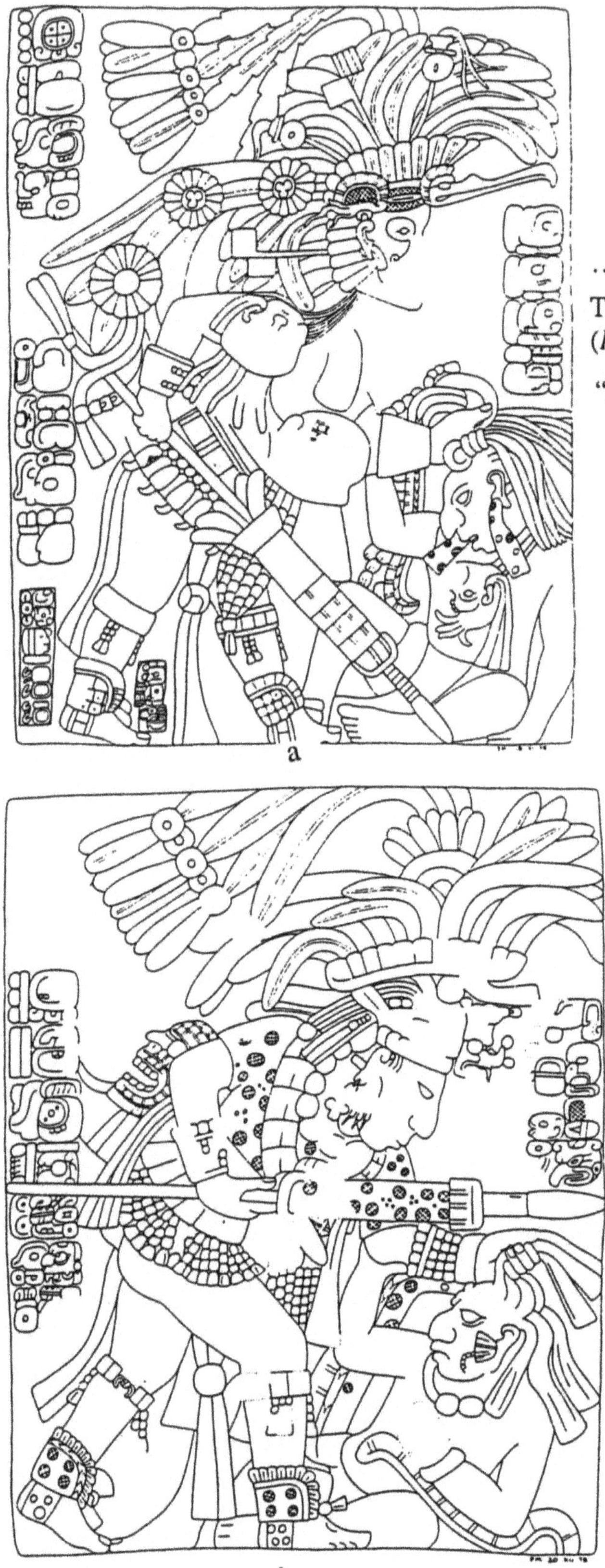

Figure 5-19 Capture of Sak Tz'i' lords by Chel Te of Yaxchilan (a) and Chan Muwan II of Bonampak (BPK. L. 1 and 2 drawings by Peter Mathews)

Five years later between 9.18.1.18.18 and 9.18.1.9.2 (March 28 to April 1, AD 792) we find Piedras Negras and its subsidiary site La Mar attacking Pomona. A second joint attack on Pomona took place in 9.18.3.5.19 (AD 794). This conclusive attack sealed forever Pomona's fate, for their most important nobles were dragged back to Piedras Negras. Stela 12 from Piedras Negras and Stela 3 from La Mar record these attacks portraying the Pomona captives, which are being presented to Ruler 7 of Piedras Negras (Figure 5-20 a and b).

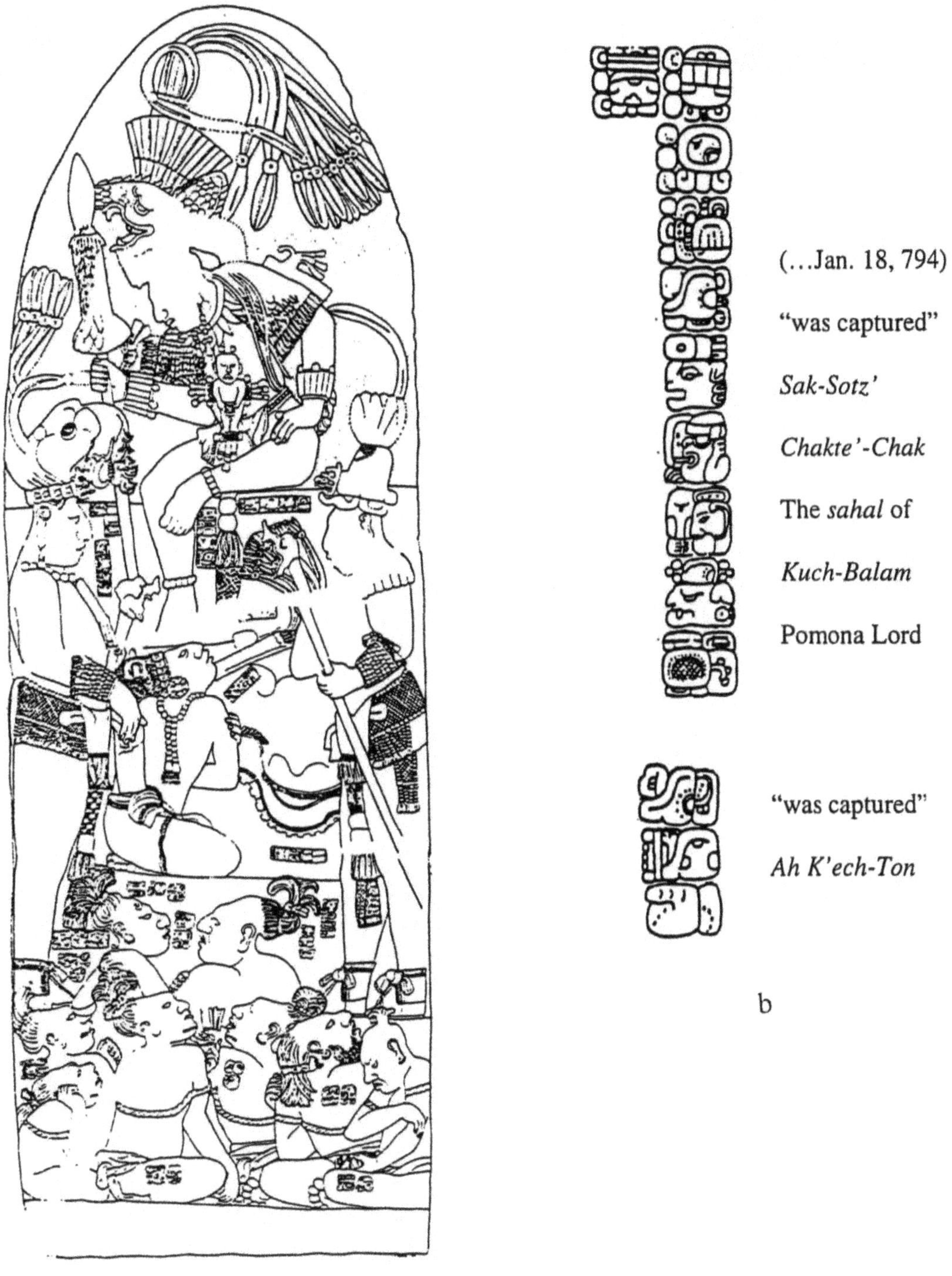

Figure 5-20 Pomona captives at Piedras Negras (a) PN Stela 12 (b) La Mar Stela 3 (drawings by Linda Schele)

The available evidence indicates that after this defeat Palenque's involvement in the region ceased. Figure 5-21 shows the territorial composition for the above-described period.

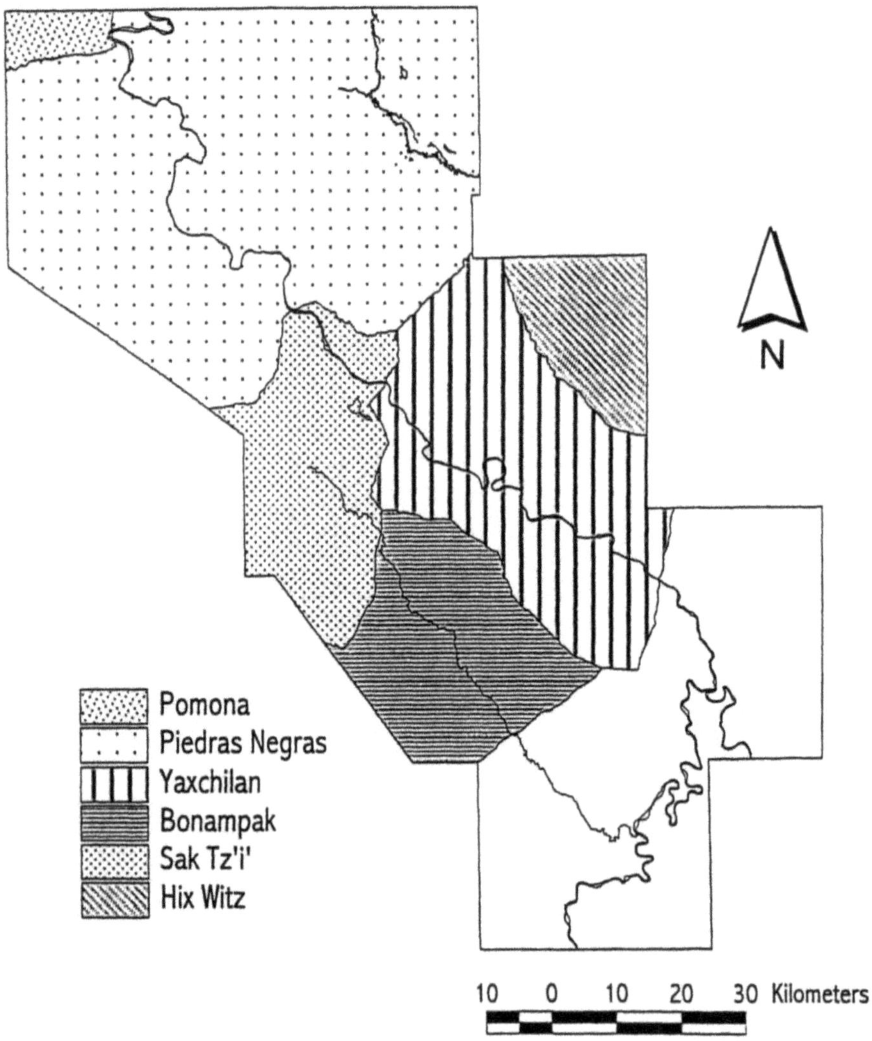

Figure 5-21 Suggested territorial extents at AD 763-772

The Final Years: 9.18.6.4.19 to 10.1.14.9.17 (AD 796-864)

The victories of Yaxchilan, Bonampak and Piedras Negras, although decisive, were short-lived. The period that followed has been characterised by Mathews as that of "endemic warfare" (Mathews 1995). From the amount of captive references that occur in Yaxchilan during Chel-Te's reign it becomes apparent that he had to spend considerable resources trying to keep all possible contestants or rebellious nobles at bay. Piedras Negras also seems to have gradually lost control of its newly recovered domain. By 9.18.15.0.0 (AD 805) Piedras Negras' most faithful ally, La Mar, is on its own. Stela 2 of La Mar records the tun ending celebration of its ruler Mo' in the company of a group of subordinates, but no mention of Piedras Negras.

Of the relevant actors in this geopolitical scenario Palenque would be the first one to go. The last inscription known for this centre is contained in a pot that has the date 9.18.9.4.4 (AD 799) it makes reference to a later ruler, namesake to the great Hanab Pakal, perhaps as an evocation of the glorious times now forever gone.

Apparently by the end of Piedras Negras' history Yaxchilan had the final say. Houston has noted that the name of the last important captive recorded in Yaxchilan's Lintel 10 bears great resemblance with that of Piedras Negras Ruler

7, the last known king of this polity. The dates are in accordance, the last known reference for Ruler 7 falls in AD 795 while the date of the capture recorded in Lintel 10 falls in AD 808 (Houston et al. 1998:10). Furthermore, Houston also notes the higher frequency of projectile points that were found around the Acropolis of Piedras Negras suggesting that, as in the case of Dos Pilas, this was where the Piedras Negras forces made their last stand. Bonampak's exuberant heir designation celebration depicted in Room 1 of Structure 1 of Bonampak seems to have fallen short of its expectations. The available evidence indicates that Chan Muwan's son never acceded to the Bonampak throne. The last known inscription for Bonampak comes from this building and it is dated to 9.18.15.0.0 (July 20, AD 805).

Mathews (1988) made special mention of the "sloppiness" in the execution of Yaxchilan's Lintel 10. This monument belonged to the last known king of Yaxchilan K'inich Tab-Kimi and as Mathews observed it appears to be the monument that represented the end the royal line at Yaxchilan. The last known date for Yaxchilan is contained in this monument: 9.18.17.13.14 (April 9, AD 808) the day Ruler 7 of Piedras Negras was captured (Mathews 1988:323-324).

As in the case of Piedras Negras Yaxchilan seems to have made its last stand at the "Little Acropolis" judging from the large number of projectile points recovered from that part of the city (Kaneko 1996 personal communication).

Paradoxically the last surviving polity of the region seems to have been Sak Tz'i'. A hieroglyphic text dating from 10.1.14.9.17 (March 29, AD 864) is contained in the Randall Stela, a funerary monument for Balam-Chilkay, sahal of K'ab'-Chan-Te' of Sak Tz'i'. This sole mention of a hierarchical relationship reflects that the political structure for this polity was still in place. However it is important to note that the sculptural style of this monument seems conspicuously foreign (Figure 5-22). It becomes apparent that after decades of fierce warring these polities were debilitated to the point that their political infrastructure literally crumbled beneath their feet, thus setting the stage for the Classic Maya Collapse in the region.

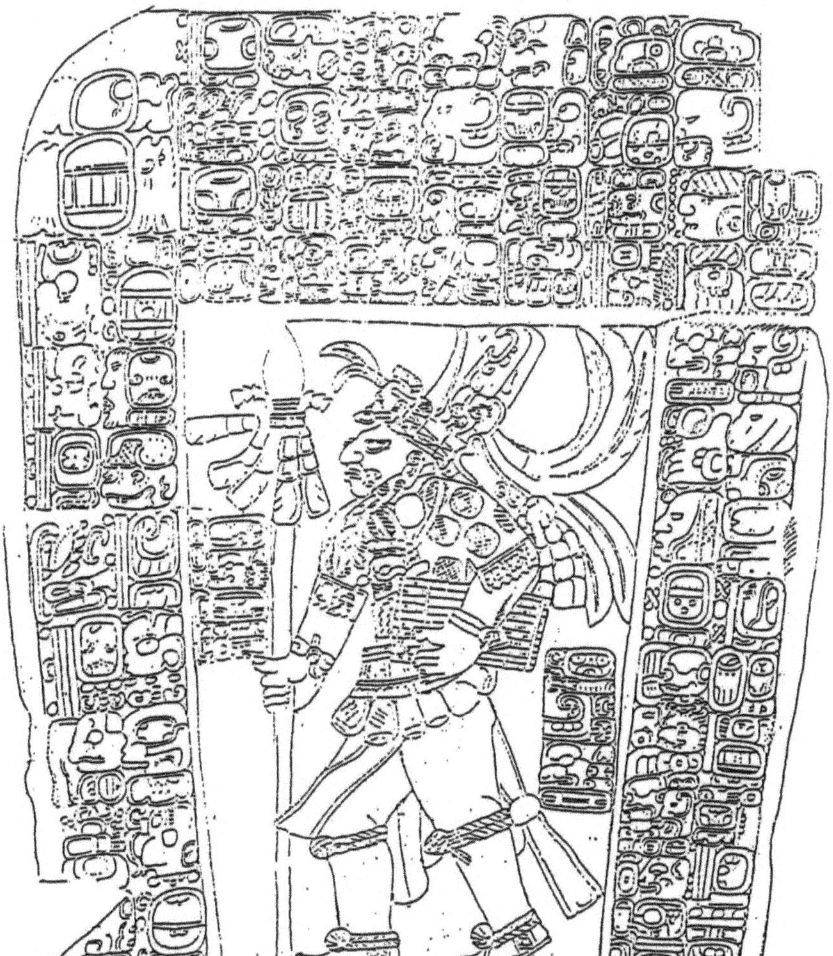

Figure 5-22 The Randall Stela, Sak Tzi'i's latest known monument (Drawing by Ian Graham)

SIX CONCLUSIONS

Site Interaction and Settlement Pattern

In the second part of the preceding chapter I offered a reconstruction of the political geography of the Upper Usumacinta region as defined in Chapter Three. This reconstruction was based on a conjunctive approach taking advantage of the historic information contained in the inscribed monuments of the region, and the potential that GIS offers to spatial modelling and analysis, within the context of Landscape Archaeology. Through this approach I have demonstrated how events occurring outside the region do in fact alter the internal composition of the political organisation within a region.

After the initial application of the Gravity Model my spatial analysis centred more on trying to define the structure of the settlement pattern in relation to other centres, rather than the internal composition within centres. In this sense the approach I use has greater affinity with Nearest Neighbour Analysis than with Central Place Theory. However the application of the Dempster-Shafer logic to the spatial data yielded information that seemed more akin to Central Place Theory in the sense that it highlighted the areas of strategic site location from the perspective of Christaller's K= 4 Transport Principle (Figure 1-1, see also Chapter One).

These results also suggested to me that the more traditional applications of the feudal model do not properly reflect the internal structure of a centre and its subsidiary sites. By this I mean that envisioning the secondary, tertiary, quaternary and so on centres as "scaled-down" versions of the primary centre, where the local nobility held court, prevents us from understanding the complexity of the interaction between these sites and their central place, as well as between the central places themselves.

The interpretation of the epigraphic evidence in light of the above-mentioned results suggests an organic type of internal polity composition. In this organic composition some subsidiary sites will have completely unique functions different from those conducted at the main centres. Aliphat concluded that sites like El Cayo, El Chile, El Chicozapote, La Pasadita and Texcoco played a prominent role in controlling access to communication routes. In a context of endemic warfare the lack of formal roads as defined by Trombold (see Chapter Five) meant that military outposts needed to be distributed across the hinterland at these strategic points.

El Cayo for example, was a secondary centre that is located at the crossroads of almost all "roads" that go to Piedras Negras. I argued in Chapter Five that this site was intensely contested by Piedras Negras and Sak Tz'i', hence it represents a location where one would expect to find evidence of a military outpost or other defensive works. However, the surveys of the area around this site carried out during the 1992 and 1993 field seasons of the Proyecto Arqueológico El Cayo, did not yield any indication of defensive works or any potential military constructions (Mathews and Aliphat 1993). Granted, only the Mexican side of the site was surveyed and there is clear evidence of constructions of considerable size on the Guatemalan side. However it is only the Mexican side of El Cayo where we find elevations close enough to the site that would offer shelter for defensive purposes.

A few hundred meters upstream from El Cayo the Usumacinta River narrows considerably and is surrounded by steep cliffs. This location would be appropriate for defensive purposes, but only in those cases where the river was used as the main means of transportation, and now we know that there were alternative inland routes. If not military, then what was the relevance of El Cayo for the internal composition of Piedras Negras or Sak Tz'i'? I think that the answer to this question lies in the physical and social dimensions of El Cayo's location.

El Cayo, as mentioned in Chapter Three offers a good landing beach virtually all year-round. This is of prime importance for watercraft transportation. It is also the point where several of the inland routes converge. Hassig has noted that roads play a very important role in collapsing social space. That these will effectively draw distant places closer or socially remove those places that are not connected (Hassig 1991:18). In the context of site interaction El Cayo seems to have functioned as a social catalyst, a place where individuals from different places converged and interacted. While Piedras Negras was the centre of political power and ceremonial activity, El Cayo may have functioned as some sort of inter-polity port of entry.

Yaxchilan's Lintels 11, 49, 37 and 35, and Piedras Negras Lintels 2 and 3 bear witness of the enormous importance that was given to foreign relations. It is unlikely, however, that contact between foreign emissaries and the local king was unmediated. Certainly some sort of formal protocol took place before the king and his royal visitors met. In this context El Cayo, which is the most important site downstream from Yaxchilan before Piedras Negras, may have functioned as the place where these rites of passage from one polity to another took place. In this respect the occurrence of the "canoe" glyph in Piedras Negras Lintel 3 and the canoe related title that one of the young Lacanha ahaws depicted in Lintel 2 uses merits a closer look. This glyph also appears in Sculpture Stone 5 of Bonampak in the context of an official visit by Yaxun Balam, so it may well be that this glyph is some form of allegory similar to the *och bi* "he entered the road" sign for death.

El Cayo, again due to its location, may have had more mundane functions purposes like port of call or provincial market or as "customs" checkpoint for watercrafts and inland "caravans" passing through it. The idea of El Cayo functioning as a marketplace may appear to contradict the basic tenet of Central Place Theory, but if viewed from the organic perspective it becomes apparent that El Cayo has a better vantage point than Piedras Negras did in terms of accessibility. Thus in this sense, it carried out a vital function for the polity as a whole. In the same tenor La Mar appeared to have had a more military function as attested by its intervention in the Sak Tz'i' and Pomona campaigns.

The initial results obtained from the Nearest Neighbour Analysis indicated a dispersion pattern. For the sake of comparison, a second analysis was performed which included the three newly located centres. The **R** coefficient with these new polities increased considerably to 2.09 indicating high level of competition between centres.

Lacanha however represents an anomalous case. The average distance between the other regional centres is of 32.25 km. Lacanha is located at scarcely 5 km from Bonampak, thus increasing the stress due to competition for the resource base. Mathews has suggested that this proximity may be due to a common origin rather than to independent settlement of the sites. Bonampak was probably an offspring of Lacanha establishing its capital at the vicinity of the latter (or vice-versa) (Mathews 1999, personal communication).

It is very likely that this was the case, the evidence from the inscriptions supports the idea of a common origin, with rulers from Bonampak using both the Bonampak and Lacanha emblem glyphs amongst their titles. Nevertheless even with this common origin the problem still persists. From the inscriptions we know that towards the Late Classic Lacanha is completely absorbed by Bonampak. However the fact that Lacanha existed as an independent polity through extended periods of time before this merge remains an issue that still awaits resolution.

The only explanation I can offer for this anomaly is that Lacanha's permanence was due to extra-regional factors. It is possible that Lacanha's loyalties were split through times between its sister polity Bonampak, and some other still unidentified centre. In this respect it is revealing that Yaxchilan which throughout its history seems to have been in close relationship with Bonampak makes war at least twice to Lacanha, taking in each case a high ranking noble as captive. Also of interest is the fact that in such occasions a different polity (i.e. Buk-Tun) associated indirectly with Lacanha was also attacked (see Mathews 1988:163). But until we can obtain more information from a systematic exploration of Lacanha and its surroundings, all of this will have to remain at a speculative level.

The distribution of these capitals across the regional landscape suggests a fragile power balance --particularly between Yaxchilan and Piedras Negras-- that was repeatedly upset by the events taking place outside the region. Viewed from this global perspective the conflicts originated between the two superpowers, Tikal and Calakmul, had a ripple effect in the Upper Usumacinta region where the different actors seem to have been always alert to "catch the tide" to their advantage.

The results presented here do not necessarily invalidate Mathews' 1988 reconstruction of the political geography of the region; however, they do highlight one of the main constraints of the Thiessen Polygon method that he used, its inherently static nature. Mathews is the first to recognise the shortcomings of this method when dealing with the dynamic processes involved in the definition of polity boundaries through time. In these terms my first map of the territorial extent of the different polities identified for the region (Figure 4-13) is as static as Mathews 1988 Thiessen Polygons. These, however, are not necessarily of dubious value; as these maps are providing us with the initial framework on which to mount the more dynamic processes.

With the use of the Thiessen polygons Mathews has been able to describe through fixed periods of time, the changes that took place in the political geography of the Maya lowlands. His is still the most cited source in this respect. On my part, by determining what would be the state of natural entropy in terms of political organisation, I have been enabled to model the effects of changes in the arrangement of political alliances. In other words I have shifted my research focus from questions of what? and how? to what if?

The maps I have derived present the plausible reconstruction of the political organisation of the region through time. These, as already mentioned, are strongly based on the epigraphic accounts and the characteristics of the natural landscape. The definition of the boundaries between the different polities is of course subject to reassessment. I do not share Hammond's assertion (1991), which suggests that boundary maintenance and territorial integrity were unimportant issues to the Maya ruler (Hammond 1991:276-277). On the contrary I believe that if the patterns of site interaction and the strategic location of sites are telling us anything at all, it is that the territorial integrity of the polity and its boundary maintenance were a priority for the Maya kings.

The central tenet of this work, is that due to the restraints imposed by the physical landscape, the location of the routes of communication constituted the dominant factor for site location. From here it follows that site interaction and internal composition could be measured in terms of access to these "roads". Likewise the level of integration to one or the other hegemonic powers depended on the level of communication that a site had with one or the other of these superpowers. This realisation helps us explain why Piedras Negras and Yaxchilan acquire preponderance in the region.

Throughout the history of the region these two sites are the only two kingdoms that make explicit mention of their relationship to the two superpowers. Their geographic location along the Usumacinta and close to the eastern Peten road, seems to have determined this relationship. It is for this reason that the Piedras Negras and Yaxchilan kings emphasise their interaction (whether marriage alliances or military take-overs) with the polities located towards the east. Hix Witz, for example, was located between these centres and the road to the Peten, while Man (La Florida) is located on the San Pedro Martir River which in turn constitutes the connection to Calakmul.

From this perspective Tikal and Calakmul represent the axis around which the Classic Maya geopolitical globe rotated. But why were these superstates so important in the Maya world? Where is the essence of their relationship with their close allies and other subordinate states? Is this disproportionate relevance due merely to military strength or is something else involved?

On reviewing the importance of roads for the political organisation of the Upper Usumacinta in the light of the networks of alliances one cannot help but notice the existence of an underlying structure that gives cohesiveness to these forms of political organisation. Without trying to underscore the importance that the economic factors had in the structuring of the political organisation, it seems to me that it were these that were influenced by factors pertaining to the ideological realm rather than the other way around. In support of this argument I refer to an issue raised by Sharer (1991) regarding the definition of what was to be understood as the Classic Maya culture. Sharer decried the manner in which our conception of the Classic Maya culture was shaped taking into account only a sector of the society, the elite, and within this, only the cultural manifestations characteristic of the Central lowlands. Sharer stresses the necessity to recognise the great diversity present in the Maya lowlands as a whole in order to elaborate a more accurate definition of Maya culture (Sharer 1991:180-198).

Although I fully agree with Sharer on the issue, I believe that our "stereotyped" notion of Classic Maya Culture (see for example Berlin 1977) still represents a valuable concept. The relevance of this definition lies in the sense that it does not represent a mental construction of the archaeologist, but rather a mental construction of the actual Classic Maya elite. This conception, though, it may not have represented the *real* state of things, may be seen in the form of an aesthetic ideal. This ideal formed a mental template or mould, on which the Maya patterned their behaviour. As an aesthetic ideal it defined the parameters of the *Maya way of life*. This ideal way of life was certainly imposed upon the whole of the population whether it responded to their social realities or not. The imposition was facilitated by means of a ritual complex that *transposed* these values to more familiar concepts, much in the same way as the *American way of life* has been imposed on most of the world through the mass media.

In this context the success of Tikal and Calakmul in drawing to their political sphere smaller centres makes perfect sense. The spatial and social proximity that these lesser centres could achieve with respect to the two capitals of the Maya world, aggrandised their own regional status by emphasising their *worthiness* of partaking with *the big ones* in this ideal.

Spatial Models in Archaeology: an Evaluation

The application of GIS modelling proved to be in this case an asset for the identification of otherwise unperceived patterns of social and spatial interaction. It was not however a straightforward process. Much thought had to be given to the definition of the issues and the relevant variables that could be properly handled by these tools. From the start I have emphasised the enormous advantages that the existence of the historic accounts contained in the Maya monuments represent. I tried to be very specific in describing the different stages considered in the construction of this explanation model. These detailed account is mainly due to a perceived need to describe as thoroughly as possible the kind of problems that the archaeologist working with these type of models may encounter, and not as "cook book" presentation.

Yet another advantage that GIS based model offers for archaeological research lies in its aggregating nature. The original data set can be increased, or edited as new relevant information is obtained, without having to re-elaborate the whole database. Especially important is the fact that GIS has the ability to integrate disparate data from different sources into a workable format. This is of particular relevance in problems related to site location or site prediction as remote-sensing imagery is becoming increasingly available.

In the context of this research the classification of the satellite and radar imagery available for the study region promises interesting results and will surely enough help me reassess movement over natural terrain. Initial probes were done with a SAR-C Radar image for the area. NASA through its Jet Propulsion Lab (JPL) kindly provided this image. These initial efforts are centring in the identification of the texture patterns around the known-site locations in order to compare them to the potential areas defined by the Gravity Model and Dempster-Shafer. The preliminary results although encouraging are still too tenuous to present in this work.

This however highlights the advantage of being able to aggregate new data in the working model in order to fine-tune it. The potential uses of these applications are even more evident in geoarchaeological problems.

For example, through the combined use of GIS the hieroglyphic inscriptions, the petrographic composition of the monuments and remotely sensed data, a model to reconstruct the quarrying patterns of limestone may be proposed. Good quality limestone must have been of tremendous importance to the Maya kings that sought perpetuating their histories in the inscribed monuments and architectural works. In this sense, through the application of a model that takes account of the location of quarries, access and control of such locations and the distribution networks of the raw materials could offer more insight of the internal composition of the polities as well as on the trade networks.

Will this model work? I believe so. However a very important part is still pending: the ground "proofing". So far it may appear that I have "let my fingers do the walking". This is not the case, what I have done is compiled the available data from dozens of specialists that have done the actual survey of the area (including myself), and distilled from these data the information relevant to my research problem.

The application of a conjunctive approach in the elaboration of this spatial model has helped me recognise some patterns in site interaction that otherwise would have been difficult to perceive. However, I am aware that this model is just a heuristic tool and as such it has inherent limitations that will become apparent when contrasting it to the real world. In this sense loyal to the tradition in Mexican archaeology of giving prevalence to fieldwork, I am still a firm believer in the maxim of the great Pedro Armillas: *La arqueología se hace caminando* "Archaeology is done by walking".

APPENDIX 1:CARTOGRAPHIC MODELS

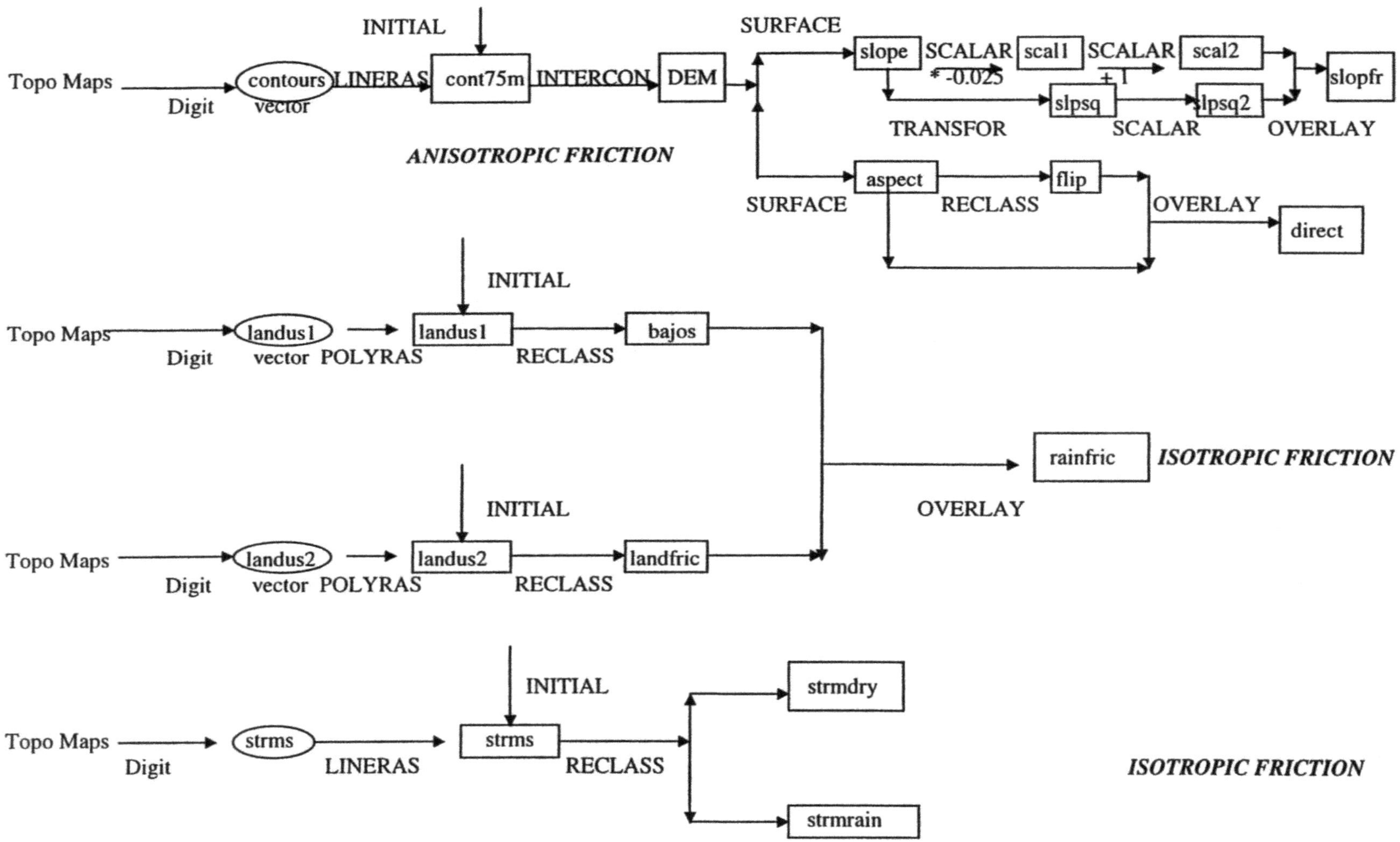

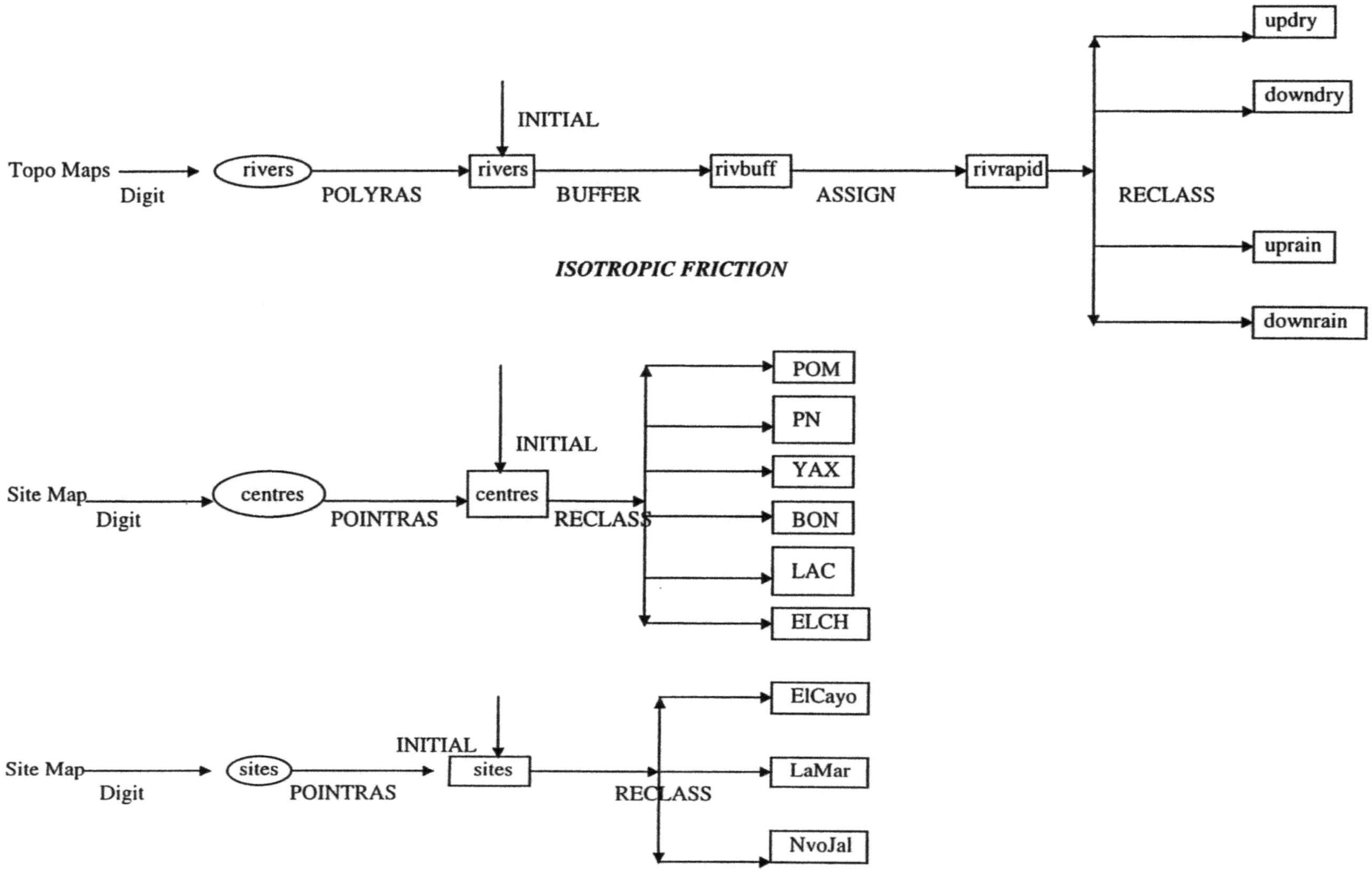
Topo Maps
Digit
rivers
POLYRAS
INITIAL
rivers
BUFFER
rivbuff
ASSIGN
rivrapid
RECLASS
updry
downdry
uprain
downrain
ISOTROPIC FRICTION
Site Map
Digit
centres
POINTRAS
INITIAL
centres
RECLASS
POM
PN
YAX
BON
LAC
ELCH
Site Map
Digit
sites
POINTRAS
INITIAL
sites
RECLASS
ElCayo
LaMar
NvoJal

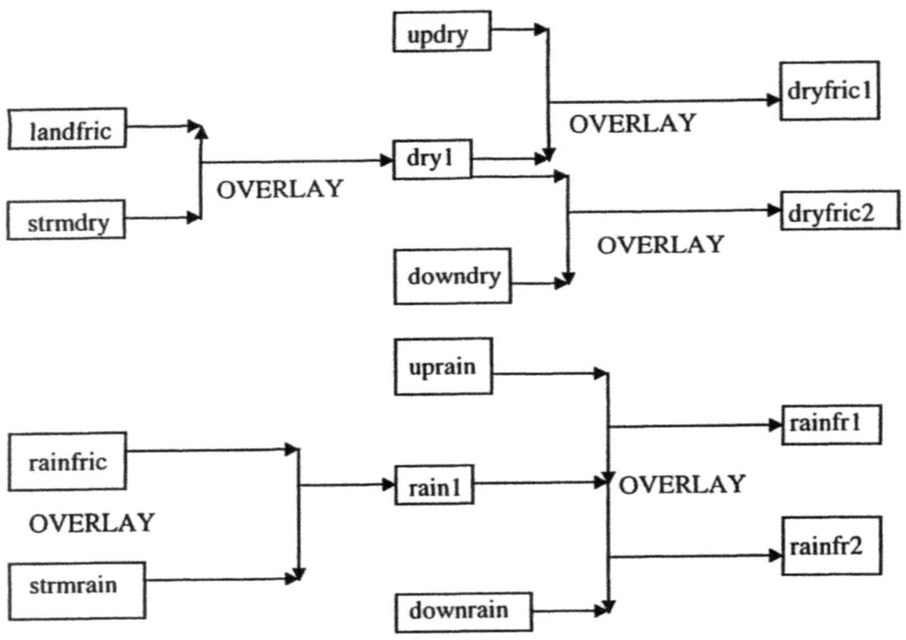

ISOTROPIC FRICTIONS

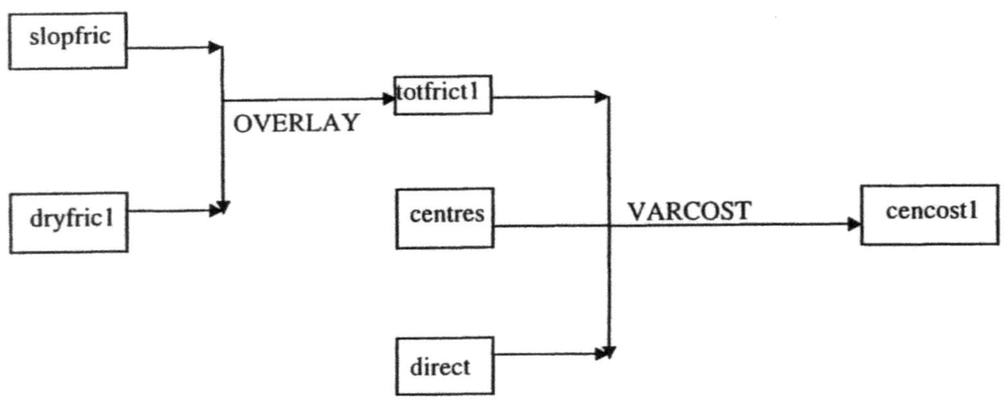

COST SURFACE DRY-DOWNSTREAM

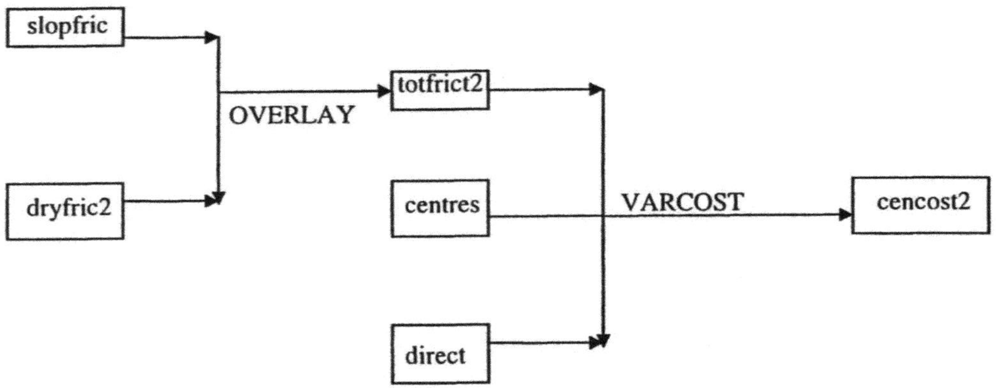

COST SURFACE DRY/UPSTREAM

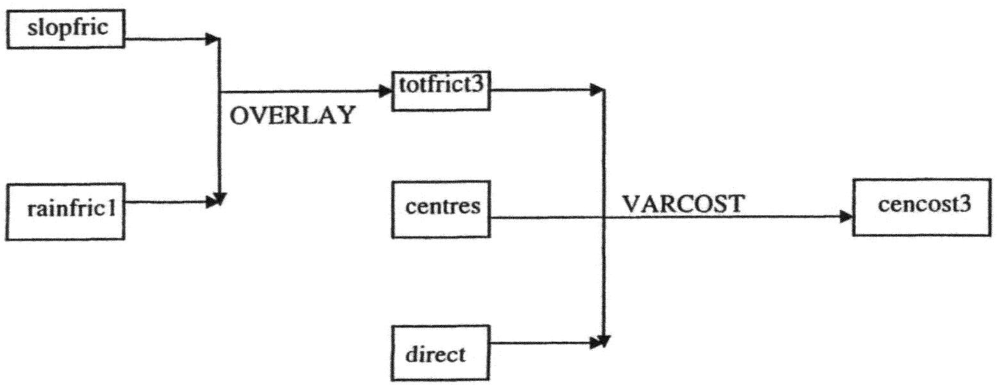

COST SURFACE RAIN/DOWNSTREAM

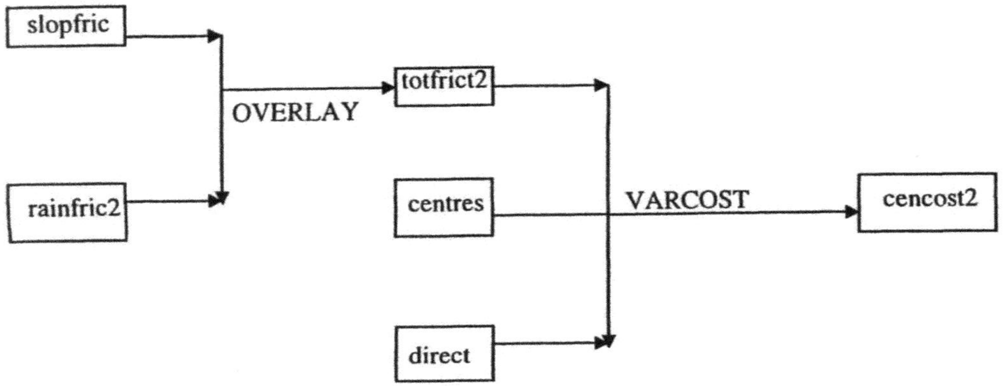

COST SURFACE RAIN/UPSTREAM

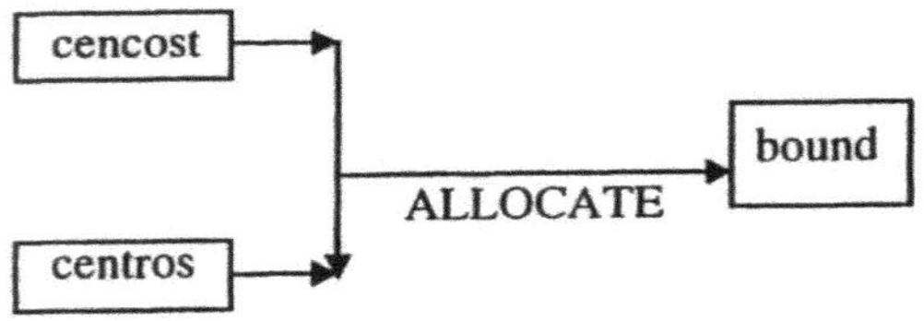

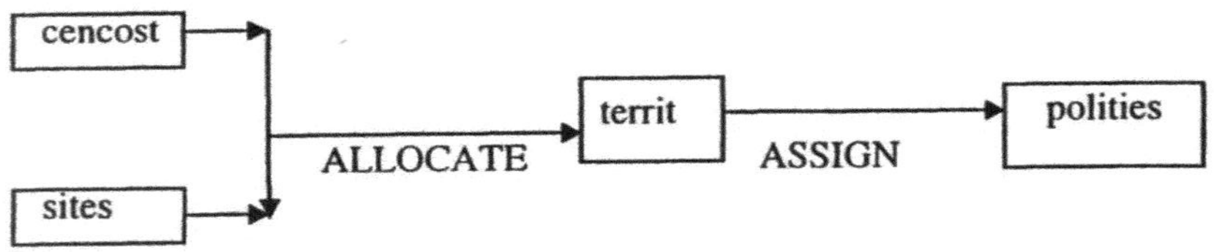

POLITY TERRITORIAL EXTENSION

APPENDIX 2: SUMMARY STATISTICS FOR THE GRAVITY MODEL

Linear Regression of the Data SPSS

Listwise Deletion of Missing Data

Equation Number 1 Dependent Variable.. I

Block Number 1. Method: Enter P1P2DIJ

Variable (s) Entered on Step One
1.. P1P2DIJ

Multiple R	.56949
R Square	.32431
Adjusted R Square	.30018
Standard Error	208.42607

Analysis of Variance

	DF	Sum of Squares	Mean Squares
Regression	1	583826.68890	583826.68890
Residual	28	1216359.97776	43441.42778

F = 13.43940 Signif F = .0010

----------------------------------Variables in the Equation--

Variable	B	SE B	Beta	T	Sig T
PIP2DIJ	159.903713	43.618289	.569486	3.666	.0010
(Constant)	251.670911	40.068267		6.281	.0000

End Block Number 1 All requested variables entered.

SITES	P1	P2	I	Dij (km)	k	dij
PN/YAX	7	3	810	45	251.6709	0.01015
PN/BPK	1	1	80	56	251.6709	0.007048
PN/LAC	1	1	80	54	251.6709	0.007048
PN/PAL	2	5	450	86	251.6709	0.009397
PN/POM	1	1	90	48	251.6709	0.006644
PN/LaM	8	2	690	14	251.6709	0.009599
PN/EC	1	5	510	11	251.6709	0.006241
YAX/BPK	3	5	655	22	251.6709	0.009539
YAX/LAC	2	1	125	25	251.6709	0.007973
YAX/LaP	1	2	200	15	251.6709	0.006304
BPK/LAC	8	3	850	5	251.6709	0.10592
BPK/LaM	1	1	60	48	251.6709	0.00601
BPK/NvoJ	1	1	100	13	251.6709	0.006304
LAC/NvoJ	1	1	90	10	251.6709	0.006644
YAX/LAXT	1	2	200	15	251.6709	0.006304
SAK/PN	4	1	440	13	251.6709	0.00601
SAK/YAX	1	1	60	25	251.6709	0.008138
SAK/BPK	3	2	360	25	251.6709	0.008138
SAK/LAC	1	1	100	11	251.6709	0.006304
SAK/LaM	2	1	180	13	251.6709	0.006644
SAK/EC	1	4	400	11	251.6709	0.006304
SAK/NvoJ	1	1	100	11	251.6709	0.006304
SAK/K-S	2	1	120	25	251.6709	0.007973
MAN/PN	1	10	600	54	251.6709	0.008138
MAN/YAX	1	5	300	54	251.6709	0.008138
IXW/PN	1	3	190	49	251.6709	0.007921
IXW/YAX	1	9	560	25	251.6709	0.007991
K-S/BPK	1	2	230	14	251.6709	0.005878
K-S/LAC	1	2	300	13	251.6709	0.005147
K-S/NvoJ	1	1	90	13	251.6709	0.005147
K-S/ECH	1	1	40	35	251.6709	0.009967
SR/YAX	5	1	500	15	251.6709	0.006304
LAK/YAX	1	2	110	25	251.6709	0.0085
BUK/YAX	1	2	120	25	251.6709	0.008138
WAK/YAX	1	1	60	25	251.6709	0.008138

APPENDIX 3 MAJOR HISTORICAL EVENTS OF THE UPPER USUMACINTA REGION

Long Count	Julian Date	Site	Event	Other Sites	Monument
8.14.2.17.6	8/2/320	Yaxchilan	Accession of Yoat-Balam		Yax. L. 11
8.16.10.0.0 *	12/12/366	Yaxchilan	Itzam-Balam I (Deity Jaguar) is ruling Yaxchilan and receives Royal visit		Yax. L. 11
8.17.1.17.16	10/6/378	Yaxchilan	Accession of Yaxun Balam I		
8.17.2.11.5	5/29/379	Yaxchilan	Yaxun-Balam I is ruling Yaxchilan and receives Royal visit.		Yax. L. 11
8.17.13.3.8	10/20/389	Yaxchilan	Yax-Deer Antler-Chami is ruling over Yaxchilan and receives Royal visit	Diving-Bird Site	Yax. L. 11
8.18.6.5.13	9/27/402	Yaxchilan	Ruler 5 is ruling over Yaxchilan and receives Royal visitor		Yax. L. 49
8.19.0.0.0 *	3/24/416	Yaxchilan	K'inich-Tab Chami I (Mahk'ina Skull I) is ruling over Yaxchilan	Bonampak	Yax. L. 49
8.19.7.11.8??	10/2/423	Yaxchilan	Moon-Chami is ruling over Yaxchilan	Piedras Negras	Yax. L. 49
9.1.0.0.0 *	8/27/455	Yaxchilan	Yaxun-Balam II is ruling over Yaxchilan	Piedras Negras	Yax. L. 37
9.3.0.0.0	1/29/495	Bonampak	Period ending celebration performed by Kan-Batz and Yak'ak, Bonampak king		Houston Panel

Long Count	Julian Date	Site	Event	Other Sites	Monument
9.3.0.14.13	11/18/495	Bonampak	Flint deposit ritual performed by Yak'ak and/or Kan-Batz at Bonampak		Houston Panel
9.3.3.16.4	12/3/498	Bonampak	The House dedication of Kan Batz *anabil* of Yak'ak Bonampak king		Huston Panel
9.3.10.0.0 *	12/7/504	Yaxchilan	Tab-Balam I is ruling Yaxchilan and receives Royal visitors	Bonampak	Yax. L. 37
9.3.13.12.19	8/7/508	Yaxchilan	Tab-Balam I is ruling Yaxchilan and receives Royal visitor	Tikal	Yax. L. 37
9.3.19.12.12	6/30/514	Piedras Negras	Tab Balam of Yaxchilan appears as subservient to the PN ruler	Yaxchilan	PN L.12
9.4.8.14.9	6/20/523	Bonampak	House dedication Waybil (house of dreams) of K'an Batz yanabil Yak'ak of Bonampak		Poh Panel
9.4.11.8.16	2/11/526	Yaxchilan	Accession of K'inich Tab Chami II		Yax. L.35,L.48, and L.38
9.4.11.8.16 *	2/11/526	Yaxchilan	K'inich-Tab-Chami II receives Royal visit byAk-Kuchah ?? Balam	Diving-Bird Site	Yax. L. 35
9.4.11.8.16 *	2/11/526	Yaxchilan	K'inich-Tab-Chami II receives Royal visit by 9 Ok'a	Lakamtun	Yax. L..35
9.4.11.8.16 *	2/11/526	Yaxchilan	K'inich- Chami II receives Royal visit from Bonampak	Bonampak	Yax. L. 35

Long Count	Julian Date	Site	Event	Other Sites	Monument
9.5.2.10.6	1/14/537	Yaxchilan	K'inich-Chami II of Yaxchilan receives a Royal visit.	Calakmul	Yax. L. 35
9.6.10.13.17	1/4/565	Sak Tz'i'(?)	A house dedication event with the possible intervention of U-K'ab Sak Tsi Ahaw		New York Panel
9.6.10.14.15	11/17/564	Yaxchilan	Tab-Balam II captures Tok-Xun (Flint Bat) of Lacanha	Lacanha	Yax. HS. 3-I
9.8.0.0.0	8/22/593	Lacanha	Period Ending by local ruler		Lac. St.1
9.8.5.13.8	4/21/599	Palenque	Calakmul wages war in Palenque	Calakmul	Pal. H.S.
9.8.6.13.17	4/24/600	Bonampak	Accession of Chan Muwan I	Yaxchilan Lacanha	Bon. S.S.4
9.8.9.15.11	5/13/603	Bonampak	War waged against Palenque, the tok' pakal was thrown down.	Palenque	Bon. L. 4
9.8.10.6.16	11/14/603	Piedras Negras	Accession of Ruler 1		St.25
9.8.12.2.9	8/8/605	Bonampak	A seating event of unclear nature	Lacanha	Bon. S.S.4
9.8.17.15.0	3/21/611	Bonampak	Capture of a Lacanha person by Chan Muwan I of Bonampak.	Lacanha	Bon. S.S.4
9.8.17.15.14	4/4/611	Palenque	Ch'akah (war event) against Palenque's capital (Lakamha) by Une-Kan Calakmul king	Pomona	Pal. Temple of the Inscriptions
9.9.1.7.1	9/22/614	Bonampak	Death of Lacanha Ruler	Lacanha	Bon. S.S.4

Long Count	Julian Date	Site	Event	Other Sites	Monument
9.9.0.0.0	5/9/613	Yaxchilan	The 14th(?) Yaxchilan Ruler appears performing an unknown event.		Yax. St. 2
9.9.11.12.3	11/11/624	Piedras Negras	Ruler 1 takes as captives K'ab' Chan Te', Sak Tz'i' ahaw and Ch'ok Balam, a Yakun of the Ch'ul ahaw of Palenque	Sak Tz'i' Palenque	PN St.26
9.9.16.10.13	9/15/629	Yaxchilan	Accession of Yaxun-Balam III		Yax. St. 6
9.10.6.2.1	2/3/639	Piedras Negras	Death of Ruler 1		PN L.1
9.10.6.5.9	4/12/639	Piedras Negras	Ruler 2 accedes		
9.10.8.3.5	2/16/641	Sak Tz'i'(?)	K'ab Chan-Te', Sak Tz'i' ahaw did something (U-kabhi) to Ruler 1 of P.N. (probably a war associated event)	Piedras Negras	Denver Panel
9.10.8.6.2	4/14/641	Sak Tz'i'(?)	Nik- Mo' of the Rabbit-Stone place, sets fire to the "seat" of K'ab' Chan Te' at Sak Tz'i'	La Mar	Denver Panel
9.10.8.6.3	4/15/641	Sak Tz'i'(?)	K'ab' Chan Te' "axes" something or someone at the Rabbit Stone place of Nik Mo'	La Mar	Denver/Brussels Panels
9.10.8.6.2	4/14/641	Sak Tz'i'(?)	Nik- Mo' of the Rabbit-Stone place, sets fire to the "seat" of K'ab' Chan' Te' at Sak Tz'i'	La Mar Piedras Negras	Denver Panel
9.10.8.6.5	4/17/641	Sak Tz'i'	K'ab' Chan Te' captures Ek Mo', Bonmapak ahaw	Bonampak	Brussels Panel

Long Count	Julian Date	Site	Event	Other Sites	Monument
9.10.14.13.0	7/30/647	Yaxchilan	Yaxun-Balam III captures Xupib (?) of Hix Witz	Hix Witz	Yax. H.S.3
9.11.6.2.1	10/21/658	Piedras Negras	Royal visit of Yaxchilan young Ahaws to the court of Ruler 2 of P.N.	Yaxchilan Bonampak Lacanha	PN L. 2
9.11.6.16.11	8/7/659	Palenque	Palenque captures Pomona lord.	Pomona	Pal. H.S.
9.11.6.16.11	7/8/659	Tikal	Capture of the war palanquin of Itzam Balam III by Nu Bak Chak of Tikal	Yaxchilan	Pal. House C
9.11.6.16.17	7/14/659	Palenque	Nu-Bak-Chak Ch'ul ahaw of Tikal arrives in the company of Hanab Pakal to Palenque	Tikal	Pal. House C
9.11.9.15.19	7/10/662	Dos Pilas	Balah Kan K'awil, Dos Pilas king and kalaw balam ahaw wage war, and the battle takes place at the Tied-Hair place	Knot-Site	DP H.S. 2
9.11.10.16.17	7/13/663	Palenque	Death of Pomona Lord at Palenque	Pomona	Pal. H.S.
9.11.16.11.6	2/3/669	Piedras Negras	War against El Cayo ?, Och-Nal K'utim, Yax-Nil	El Cayo	PN St.37
9.12.5.7.0	10/20/677	Yaxchilan	Capture done by Itzam-Balam III		Yax. L. 44
9.12.8.14.1	2/22/681	Yaxchilan	Capture of Ah-Nik, by Itzam Balam II	Man	Yax. H.S.3-III; L. 45; St.18
9.12.9.8.1	10/20/681	Yaxchilan	Accession of Itzam Balam III		Yax. H.S.3-III; L. 25
9.12.11.5.18	8/28/683	Palenque	Hanab-Pakal dies		Pal. Sarcophagus

Long Count	**Julian Date**	**Site**	**Event**	**Other Sites**	**Monument**
9.12.11.6.9	9/8/683	Bonampak	Accession of Ah-Na-Chuy to Bonampak ahawship		Bon. S.S.1
9.12.11.12.10	1/7/684	Palenque	Kan-Balam of Palenque accedes		Pal. Palace Tablet
9.12.13.4.3	7/13/685	Piedras Negras	Kohal headress event for PN Ruler 2 U-kabhi a Calakmul subordinate	Calakmul	Hellmuth Panel
9.12.14.10.14	11/16/686	Piedras Negras	Death of Ruler 2		PN St.8
9.12.14.13.1	1/2/687	Piedras Negras	Accession of Ruler 3		PN St.3, St.8
9.13.0.0.0	15/03/692	El Chorro	Mention of Lady from	Knot-Site	EC Altar 6
9.13.5.12.13	10/28/697	Yaxchilan	Unknown event between Itzam Balam II and one of his wives Lady Ik Kimi	Calakmul	Yax. L. 53
9.13.6.5.11	6/3/698	Yaxchilan	Itzam Balam III takes a captive		Yax. H.S.3-VI
9.13.9.14.14	11/17/701	Yaxchilan	Capture by Itzam Balam III of Tun Wob Ah K'in Ush		Yax. St.20
9.13.10.1.5	2/16/702	Palenque	Kan Balam dies		
9.13.10.6.8	5/30/702	Palenque	K'an-Hok-Chitam acceds to the Palenque kingship		Pal. Palace Tablet
9.13.17.15.16	10/28/709	Yaxchilan	Blood letting ritual performed by Lady Xoc		Yax. L. 24
9.13.18.8.0	5/20/710	Yaxchilan	Itzam Balam II takes captive		Yax. H.S.3-VI

Long Count	Julian Date	Site	Event	Other Sites	Monument
9.13.19.13.3	8/26/711	Tonina	K'an-Hok-Chitam of Palenque appears as captive at Tonina	Palenque	Ton. Mon. 122
9.14.1.17.14	11/14/713	Yaxchilan	Itzam Balam III captures Ah K'an U-Sih Baktun ahaw	Buk-Tun	Yax. L.46
9.14.3.8.4	4/28/715	Bonampak	Etz'nab Ch'oy of Bonampak declares himself yahaw of Bak Nal Chak (Ruler 3) of Tonina	Tonina	St. Louis Hieroglyphic Column
9.14.10.4.2	12/30/721	Palenque	Accession of Akul Anab		Pal. Tablet of the Slaves
9.14.10.5.0	1/17/722	Nuevo Jalisco(?)	Accession date of Tab Balam at Lacanha U-kabhi K'ab-Chan Te' Sak Tz'i' ahaw	Sak Tz'i' Bonampak Lacanha	NJ Panel 1, 2
9.14.11.4.14 *	1/7/723	Dos Pilas	Itzam K'awil (Ruler 2) captures an ahaw from Yaxchilan	Yaxchilan	DP H.S.1-III
9.14.12.7.2 *	2/18/724	Piedras Negras	3 K'atun birthday of Ruler 3		PN St.8
9.14.13.10.8	4/19/725	Yaxchilan	Itzam Balam III is attired for war by Lady Xoc		Yax. L. 26
9.14.13.10.8	4/19/725	Site R	Itzam Balam III is attired for war and performing pre-war rituals in the company of his sahals	Yaxchilan	Yax. L.1 & L. 2
9.14.15.0.0	9/16/726	Sak Tz'i'(?)	A house dedication event performed by K'ab-Chante Ch'ul Ahaw of Bonampak and Sak Tz'i'	Bonampak	New York City Panel

Long Count	Julian Date	Site	Event	Other Sites	Monument
9.14.15.0.0?	9/16/726	Lacanha	PE connected to an earlier date involving yahawte K'inich Chak-Chih, Lacanha ahaw, yahaw of K'ab' Chan Te' Sak Tz'i' and BPK holy lord	Bonampak Sak Tz'i'	Zurich
9.14.17.15.11	7/10729	Yaxchilan	Capture of Pol-Chay Ah Pay Mo' (Chuwen) by Itzam Balam III	Lacanha	Yax. H.S.3-I
9.14.18.3.13	11/9/729	Piedras Negras	Accession of Ruler 4		PN Alt.2
9.14.18.14.13	6/17/730	Tonina	Tomb dedication of K'inich Baknal-Chak (Ruler3)		Tonina Disk
9.14.19.14.18	6/17/731	Yaxchilan	An ahaw from Hix Witz performs a conjuring rite	Hix Witz	Yax. H.S.3
9.15.0.12.0	4/14/732	Yaxchilan	Capture of Nak-Chiu, Hix Witz ahaw by Itzam Balam III	Hix Witz	Yax. H.S.3-V
9.15.9.3.14	9/13/740	Bonampak	Capture of Cha vassal of the Bonampak/Lacanha lord Tab Balam (he who is from the knot place), U-kabhi Ah Sak-T'el-Huh of Lacanha/Bonampak?	Lacanha Knot-Site	Bon. L. 3
9.15.10.0.0 *	6/26/741	Copan	Somewhere around this date Smoke Shell of Copan established a marriage alliance with Palenque	Palenque	Cop. St.8

Long Count	Julian Date	Site	Event	Other Sites	Monument
9.15.10.0.1	6/27/741	Yaxchilan	Vision Serpent ritual by Yaxun Balam IV, Lady Ik-Kimi, Lady Chak-Chami, and Lord Chak-Chami. Ub-kahiy Yukom of Calakmul	Calakmul	Yax. L. 39,L. 35, & L.14
9.15.10.17.14	6/15/742	Yaxchilan	Death of Itzam Balam III		Yax. St. 12; Alt.1
9.15.11.17.3	6/15/742	Lacanha	Seating in *sahal*ship of Ah-Sac-Tel-Huh	Bonampak Knot-Site	Lac. L.1
9.15.11.17.3	6/15/742	Lacanha	Seating in *sahal*ship of Ah-Sac-Tel-Huh	Knot-Site	Lac. L.1
9.15.13.6.9	10/17/744	Yaxchilan	Yaxun Balam IV played a ball game and sacrificed Ek'-Chih a Lakamtun ahaw	Lakamtun	Yax. H.S.2
9.15.16.6.9 *	10/17/744	Dos Pilas	K'awil Chan-Kinich (Ruler 4) of DP, captures Chak-Chan-ha-Xoc Yaxchilan ahaw	Yaxchilan	D.P. H.S.3-II
9.15.15.0.0	5/31/746	Lacanha	Sak-Tel-Huh celebrates his first P.E. as sahal of Lacanha recognizing Tab Balam of BPK & LAC as his overlord.	Bonampak Knot-Site	Lac. L.1
9.15.15.0.0	5/31/746	Yaxchilan	Yaxun Balam IV performs blood letting and celebrates P.E. at Yaxchilan.		Yax. St. 11

Long Count	Julian Date	Site	Event	Other Sites	Monument
9.15.15.0.0	5/31/746	Piedras Negras	Ruler 4 of PN celebrates the PE in the company of his subordinate from the Rabbit Stone Place	La Mar	PN St.40
9.15.15.16.16	5/2/747	Bonampak	A Xuklan ahaw accedes U-kabhi Yaxun Balam IV	Yaxchilan	Bon. S.S.5
9.15.18.3.13	7/27/749	Piedras Negras	Celebration of 1st Katun of Ruler 4 of P.N. attended by Sak-Hukub Balam, of Yaxchilan (Ruler?).	Yaxchilan	PN L. 3
9.15.18.3.15	7/29/749	Piedras Negras	Ruler 4 of PN dances and drinks fermented cacao in the presence of his Yaxchilan guests.	Yaxchilan	PN L. 3
9.15.19.1.1	5/31/750	Yaxchilan	Yaxun Balam IV displays his K'awil scepter in front of three captives		Yax. St. 11
9.15.19.2.2	6/21/750	Site R	Ah Ka-Mo' does a xikbalel (war preparation rite) in front of his lord Yaxun Balam IV of Yaxchilan	Yaxchilan	Site R L.3
9.16.0.0.0	5/5/751	Yaxchilan	P.E. celebrated by Yaxun Balam IV at Yaxchilan		Yax. Alt.9; H.S.4-III
9.16.0.0.0	5/5/751	Pomona	This K'atun was ended by K'inich Ho-Ix king of Pomona in the company of a yitah of K'inich Kan-Balam Palenque Ch'ul ahaw.	Palenque	Pom. "New Stela"

Long Count	Julian Date	Site	Event	Other Sites	Monument
9.16.0.13.17	2/6/752	Yaxchilan	Capture of Kib-Tok ("Inverted Pot"), sahal of a place named Wak'ab.	Wak'ab	Yax. L. 16
9.16.0.14.5	2/14/752	Yaxchilan	Itzam Balam IV is born, Yaxun Balam IV and his wife Lady Balam Mut perform blood letting	Hix Witz	Yax. L.13; L.17
9.16.1.0.0	4/29/752	Yaxchilan	Yaxun Balam IV officially accedes to power in Yaxchilan		Yax. St.11, 12; L.1, 30; Alt.4; H.S.4-III
9.16.1.2.0	6/8/752	Yaxchilan	Yaxun Balam IV danced the xukpi with one of his wives: Lady Wak Halam-Chan ahaw of Motul de San José	Motul de San José	Yax. L.5, L.42
9.16.1.8.6	10/12/752	Yaxchilan	Yaxun Balam IV dances the chak k'at with K'an-Tok, and his wife Lady Mutul-Balam from Ix-Witz	Hix Witz	Yax. L.6, L.43
9.16.1.8.8	10/14/752	Yaxchilan	Yaxun Balam IV dances with his wife Lady Chak Kimi		Yax. L.7
9.16.1.13.17	1/3/753	Site R	Dance performed by the local sahal who recognizes subordination to Yaxun Balam IV of YAX	Yaxchilan	Looted lintel
9.16.3.16.19	3/24/755	Yaxchilan	Yaxun Balam's wife, Lady Wak-Halam-Chan ahaw of Motul de San José conjured a vision serpent	Motul de San José	Yax. L.15

Long Count	Julian Date	Site	Event	Other Sites	Monument
9.16.4.1.1	5/5/755	Yaxchilan	Yaxun Balam IV captures Jewelled Skull in the company of K'an-Tok who captured Kot-Ahaw		Yax. L.8; L.41
9.16.4.6.17	8/29/755	Yaxchilan	Dedication of the tomb of Lady K'abal-Xoc		Yax. L.28
9.16.5.0.0	4/8/756	Yaxchilan	Yaxun Balam IV dances with his subordinate K'in-Mo ahaw, 3 K'atun		Yax. L.3; L.54; L.58
9.16.6.10.19	10/21/757	Piedras Negras	Yaxun Balam IV of Yaxchilan did some action (U-kabhi) at PN, he meets with Ruler 3 of PN	Yaxchilan	PN L.3
9.16.6.11.0	11/9/757	Yaxchilan	Yaxun Balam is presented with four captives by a subordinate lord.		Yax. L.12
9.16.6.11.17	11/26/757	Piedras Negras	Ruler 4 dies		Yax. L.3
9.16.6.12.2	12/1/757	Piedras Negras	Piedras Negras records this date as the accession of Yaxun Balam IV	Yaxchilan	Yax. L.3
9.16.6.17.1	3/10/758	Piedras Negras	Ruler 5 acceded		PN St.14
9.16.7.0.0	3/29/758	Yaxchilan	Yaxun Balam's wife Lady Mutul-Balam of Hix Witz conjures the K'awil	Hix Witz	Yax. L. 40
9.16.8.3.18	6/10/759	La Pasadita	Yaxun Balam IV captures Bal-?ku and displays him in front of his sahal Tilom	Yaxchilan	LP L. 2

Long Count	Julian Date	Site	Event	Other Sites	Monument
9.16.12.5.14	6/25/763	Yaxchilan	Yaxun Balam's IV wife Lady Wak-Tun of Motul de San José conjured a K'awil	Motul de San José	Yax. L. 38
9.16.12.2.6	4/18/763	El Cayo	Sak Tz'i' lord and PN Ruler 5 are named.	Sak Tz'i' Piedras Negras	EC Pan.1
9.16.12.4.10	6/1/763	El Cayo	Ah Chak-Zotz' K'utim came out in sahalship U-kabhi Ah Sak-Max (Zotz') Sak Tz'i ahaw	Sak Tz'i' Piedras Negras	EC Pan.1
9.16.12.5.14	6/25/763	Yaxchilan	Yaxun Balam's IV wife, Lady Wak-Tun Motul de San Jose ahaw conjured a K'awil	Motul de San José	Yax. L.39
9.16.13.0.7	3/4/764	Palenque	K'uk Balam was seated in Ahwship		Pal. Tablet of the 96 Glyphs
9.16.12.10.8	9/27/763	Piedras Negras	A subordinated lord from the "Rabbit Stone" place accedes into ahawship	La Mar	PN St.16
9.16.15.0.0	2/15/766	Yaxchilan	Yaxun Balam IV and his son Chel-te K'inich (Itzam Balam IV) danced to commemorate the P.E.		Yax. L.52
9.16.15.0.0	2/15/766	La Pasadita	Yaxun Balam IV scattered in the presence of Tilom sahal of La Pasadita	Yaxchilan	LP L.1
9.16.16.0.9	2/23/767	Yaxchilan	Yaxun Balam IV adorns a captive named "Star-Sky"		Netherlands Lintel

Long Count	Julian Date	Site	Event	Other Sites	Monument
9.16.16.12.2	10/10/767	Site R	Yaxun Balam IV danced in the presence of Ah Ka-Mo'	Yaxchilan Laxtunich	Site R L.4
9.16.17.6.12	6/16/768	Yaxchilan	Yaxun Balam IV and Lord Chak Chami dance the hasaw-chan dance		Yax. L.9
9.16.18.0.19	2/18/769	Laxtunich	Fire was drilled U-kabhi Chel-te-Chan-K'inich in the company of the captor of Ba-Way	Yaxchilan	Lax. L.4
9.17.0.0.0	1/20/771	Yaxchilan	Scattering ceremony celebrating a P.E. in which chel-te (Itzam Balam IV) is involved		Yax. St.7
9.17.0.0.0	1/20/771	El Chorro	"The captor of ? ch'ah"		EC Alt.4
9.17.0.16.1	12/7/771	La Pasadita	Tilom danced with a paper decorated spear		LP L.4
9.17.1.2.12	3/7/772	El Cayo	Mention of PN k'in ahaw	Piedras Negras	EC Pan.1
9.17.1.5.9	5/5/772	El Cayo	Chan-Panak Wayib Ah Ek' Zotz' K'utim Ah Yax-nil came out in sahalship U-kabhi Ah Sak-Max Sak Tz'i ahaw	Sak Tz'i'	EC Pan.1
9.17.1.5.9	5/5/772	El Cayo	Chan-Panak Wayib Ah Ek' Zotz' K'utim Ah Yax-nil came out in sahalship U-kabhi Ah Sak-Max Sak Tz'i ahaw	Piedras Negras	EC Pan.1
9.17.2.3.7	3/17/773	Laxtunich	Chel-Te Chan-K'inich (Itzam Balam IV) is presented a headress in the company of Ka-Mo'	Yaxchilan	Lax. L.4

Long Count	Julian Date	Site	Event	Other Sites	Monument
9.17.5.8.9	6/11/776	Bonampak	Chan-Muwan II came out into ahawship at Bonampak		Bon. St. 2
9.17.10.9.4	5/31/781	Piedras Negras	Ruler 7 acceded		PN Thr.1
9.17.12.4.9	2/15/783	La Mar	The Sak Hunal was displayed for Mo' in his accession rites		LM St.1
9.17.12.13.14	8/19/783	Laxtunich	Ba-Waybi was captured U-kabhi Ah-Chak-Ma-Chami sahal		Fort Worth Lintel
9.17.14.3.8	1/14/785	Bonampak	Chan Muwan captures a prisoner		Bon. St.3
9.17.15.0.0	11/2/785	La Mar	Mo' celebrates the end of a tun		LM St.1
9.17.16.3.8	1/4/787	Bonampak	Xu???ak a yahaw of the Sak Tz'i' (?) ahaw was captured by Chel-te Chan captor of Tah Mo' Holy Yaxchilan Ahaw (Itzam Balam IV)	Yaxchilan Sak Tz'i'	Bon. L.2
9.17.16.3.12	1/8/787	Bonampak	Ah-Ho-Bak, the yahawte of Yet K'inich Sak Tz'i' ahaw is captured U-kabhi Chan Muwan II	Sak Tz'i'	Bon. L.1
9.17.16.14.19	8/23/787	Piedras Negras	Anabil Ah ?K'in K'ul Tok' was captured		PN St.12

Long Count	Julian Date	Site	Event	Other Sites	Monument
9.17.18.15.18	8/31/789	Bonampak	Blood letting rite performed by Chan Muwan's II mother Lady Akul-Patah, Lady Sahal Yaxun, and Chan Muwan's wife Lady Yax Rabbit Ch'ul Yaxchilan Ahaw	Yaxchilan Lacanha	Bon. St.2
9.18.0.0.0*	10/7/790	Tonina	Captive from Sak Tz'i' is displayed	Sak Tz'i'	Ton. Mon. 83 and looted fragment
9.18.0.3.4	12/10/790	Bonampak	Chan Muwan designates his son as his heir, event takes place "in the presence of" Chel-Te-Chan-K'inich, Ch'ul ahaw of Yaxchilan	Yaxchilan	Bon. Str.1 Room 1
9.18.1.8.18	3/28/792	La Mar	A ch'ak (war related) event was done against Pomona.	Pomona	LM St.3
9.18.1.9.2	4/1/792	Piedras Negras	War (star) event against Pakab (Pomona), which resulted in a series of captives taken.	Pomona	PN St.12
9.18.1.15.5	8/2/792	Bonampak	Battle conducted by Chan Muwan		Bon. Str.1 Room 2
9.18.3.5.19	1/18/794	Piedras Negras	War event on Pomona, Sak Sotz' and Ah K'ech are recorded as captives	La Mar	PN St.12
9.18.3.5.19	1/18/794	La Mar	Sak-Sotz' Chakte Chak sahal of Kuch-Balam of Pomona was captured, Ah-K'ech was also captured	Pomona Piedras Negras	LM St.3

Long Count	Julian Date	Site	Event	Other Sites	Monument
9.18.4.7.10	2/13/795	El Cayo	Lady Hob placed a cache U-kabhi Ah-Yax-Tuxum (?) sahal K'utim		Cleveland Panel
9.18.6.4.19	12/13/796	Yaxchilan	Ah Naman underwent some ritual after his capture by Chel-te Chan K'inich		Yax. H.S.5
9.18.9.4.4	11/13/799	Palenque	Accession of Chan-Ch'ok Tzuk, Bolon Ek'-Kab Wak Kimi Hanab-Pakal		Pal. Initial Series Pot
9.18.15.0.0	7/20/805	La Mar	Mo' the ruler of La Mar celebrates the tun ending with two or three of his subordinates.		LM St.2
9.18.17.12.6	3/12/808	Yaxchilan	Tab- Chami IV threw an unidentified object, star-shell event at K'utel Yaxha, an action was done with the flint shield of Itzam Balam IV.		Yax. L.10
10.1.14.0.14	8/28/863	Sak Tz'i'(?)	Balam-Chilkay the sahal of K'ab-Chan-Te of Sak Tz'i' died.		Randall Stela
10.1.14.9.17	3/29//864	Sak Tz'i'(?)	Smoke entered in the carved stone of Balam Chilkay, sahal (the tomb)		Randall Stela

Bibliography

Adams, R. E. W.

1981. "Settlement patterns of the central Yucatan and southern Campeche regions" in *Lowland Maya Settlement Patterns*, ed Wendy Ashmore, University of New Mexico Press, pp. 211-257.
1986. Rio Azul Reports 2: The 1984 Season. San Antonio: Center for Archaeological Research, University of Texas.

Adams, R.E.W. and R.C. Jones

1981. Spatial Patterns and Regional Growth Among Classic Maya Cities. *American Antiquity*, Vol.46, No.2, pp.301-322.

Adams, R.E.W. and W.D. Smith

1981. Feudal Models for Classic Maya Civilization. In *Lowland Maya Settlement Patterns*, ed. Wendy Ashmore, A School of American Research Book.

Aliphat, M. M.

1994. Classic *Maya Landscape in the Upper Usumacinta Valley.* Ph.D. dissertation, Department of Archaeology, University of Calgary.

Allen, K. M. S., S. W. Green and E. B .W. Zubrow

1990. "Interpreting space" in Interpreting Space: *GIS and Archaeology*, ed. Kathleen M.S. Allen, Stanton W. Green, Ezra B.W. Zubrow, Taylor and Francis, pp.383-386.

Allen, W. and J.B. Richardson

1971. "The reconstruction of kinship from archaeological data: the concepts, the methods and the feasibility". In *American Antiquity* 46, Vol. 36 no.1, pp.41-53

Ashmore, W.

1981. Some Issues of Method and Theory in Lowland Maya Settlement Archaeology. In *Lowland Maya Settlement Patterns*, ed. Wendy Ashmore, A School of American Research Book,

Ashmore, W. and G.R. Willey

1981. A Historical Introduction to the Study of Lowland Maya Settlement Patterns. In *Lowland Maya Settlement Patterns*, ed. Wendy Ashmore, A School of American Research Book,

Ball, J.W. and J. Taschek

1991. Late Classic Lowland Political Organization and Central-Place Analysis. *Ancient Mesoamerica*, Vol.2, pp.149-165.

Barthel, T. S.

1968. El Complejo Glifo Emblema. *Estudios de Cultura Maya* Vol. VII, UNAM, Mexico, pp.99-108

Berlin, H.

1958. El glifo "emblema" en las inscripciones mayas. *Journal de la Société des Americanistes* 47: pp. 111-119.

1977. Signos *y Significaciones en las Inscripciones Mayas*. Instituto Nacional del Patrimonio Cultural de Guatemala.

Berry, J. K.

1995. Spatial *Reasoning for Effective GIS.* GIS World Books, Fort Collins, Colorado

Brannan J. A.

1992. On Modelling Resource Transport Costs: Suggested Refinements. *Current Anthropology*, Vol.33, No.1, 56-59.

Bratfisch O.

1969. A Further Study of the Relation Between Subjective Distance and Emotional Involvement. *Acta Psychologica*, Vol.29, 244-255.

Bullard, W.R. Jr.

1960. Maya Settlement Pattern in Northeastern Peten, Guatemala. *American Antiquity*, Vol.25, No.3, pp.355-372.

Burrough, P. A.

1994. Principles *of Geographical Information Systems for Land Resource Assesment.* Monographs on Soil and Resources Survey No. 12, Oxford Science Publications.

Carmichael, D. L.

1990. GIS predictive modelling of prehistoric site distribution in Central Montana. In *Interpreting Space: GIS and Arhaeology*, Kathleen M. S. Allen, Stanton W. Green and Ezra B. W. Zubrow (eds.), Taylor and Francis, pp.216-225.

Charnay, D.

1887. The *Ancient Cities of the New World.* Chapman and Hall Ltd. London

Chisholm, M.

1968. Rural *Settlements and Land Use.* Hutchinson, London.

Clark Labs

1995. Idrisi for Windows version 2, User's Manual. The Idrisi Project, The Clark Labs for Cartographic Technology and Geographic Analysis, Clark Universty, Worcester MA.

Clarke, D.L.

1977. Spatial *Archaeology.* David Clarke (ed.), Academic Press

Chorley, R. J. and P. Hagget

1967. Models *in Geography.* Methuen, London.

Crumley, C. L. and W. H. Marquardt

1990. "Landscape: a unifying concept in regional analysis". In Interpreting Space*: GIS and Archaeology*, ed. Kathleen M.S. Allen, Stanton W. Green, Ezra B.W. Zubrow Taylor and Francis, pp. 73-79

de Vos, J.

1988. La *Paz de Dios y del Rey. La Conquista de la Selva Lacandona* (1525-1821). Fondo de Cultura Economica-Secretaria de Educacion y Cultura de Chiapas, Mexico.

Denevan, W.M.

1991. "Prehistoric roads and causeways of lowland tropical America" in *Ancient Road Networks and Settlement Hierarchies in the New World*, Charles D. Trombold ed. Cambridge University Press, pp.230-242

Devine J.

1985. The Versatility of Human Locomotion. *American Anthropologist*, Vol.87, 550-570.

Drennan R. D.

1984. Long-Distance Transport Costs in Pre-Hispanic Mesoamerica. *American Anthropologist*, Vol.86, 105-112.

Duggan A. and M. F. Haisman

1992. Prediction of the Metabolic Cost of Walking With and Without Loads. *Ergonomics*, Vol.35, No.4, 417-426.

Dunham, P.S.

1990. Coming *Apart at the Seams: theClassic development and demise of Maya civilization (a segmentary view from Xnaheb, Belize).* Ph.D. dissertation State University of New York.

Ebert J.I.

1984. Remote Sensing Applications in Archaeology. In *Advances in Archaeological Method and Theory*, Vol 7, ed Michael Scheffer, 293-361.

Fox, C.

1922. The *Archaeology of the Cambridge Region.* Cambridge University Press, Cambridge.

Fox, R.G.

1977. Urban *Anthropology: Cities in their Cultural Settings.* Prentice-Hall, Englewood Cliffs, NJ.

Freidel, D.A.

1981. The Political Economics of Residential Dispersion Among the Lowland Maya. In *Lowland Maya Settlement Patterns*, ed. Wendy Ashmore, A School of American Research Book,

Gaffney, V. and Stancic Z.

1991. GIS *approaches to regional analysis: A case study of the island of Hvar.* Research Institute for the Faculty of Arts and Science, University of Ljubljana

García Moll, R. and Juarez Cossío D.

1990. Yaxchilan. *Antologia de su descubrimiento y estudios.* Coleccion Cientifica, INAH-SEP, Mexico.

Giovannini, H. Aliphat M. and Tovalín A.

1998. Soils Vegetation and Settlement in Bonampak, Chiapas, Mexico. Paper presented in the 31st Annual

Chacmool Conference, University of Calgary, Calgay.

Gorenflo L. J. and N. Gale

1990. Mapping Regional Settlement in Information Space. *Journal of Anthropological Archaeology*, Vol.9, 240-274.

Green, S. W.

1990. "Approaching archaeological space: an introduction to the volume" in *Interpreting Space: GIS and Archaeology*, ed. Kathleen M.S. Allen, Stanton W. Green, Ezra B.W. Zubrow
Taylor and Francis, pp.3-8.

Grube, N. and S. Martin

1998. Deciphering Maya Politics. In *Notebook for the XXIInd Maya Hieroglyphic Forum at Texas.* University of Texas at Austin.

Guenter, S. P. and M. Zender

1999. Palenque and Yaxchilan's War against Piedras Negras. Unpublished ms.

Haggett, P.

1965. Locational *Analysis in Human Geography*. Arnold, London.

Hammond, N.

1972. "Locational models and the site of Lubaantun: a Classic Maya Centre" in *Models in Archaeology*. ed. David L. Clarke, London: Methuen, pp.757-800

1974. The Distribution of Late Classic Maya Major Ceremonial Centres in the Central Area. In *Mesoamerican Archaeology*, ed. Norman Hammond, Duckworth

1975. Lubaantun *a Classic Maya Realm.* Peabody Museum Monographs Number 2, Harvard University

1981. Settlement Patterns in Belize. In *Lowland Maya Settlement Patterns*, ed. Wendy Ashmore, A School of American Research Book,

1991. Inside the Blackbox: Defining Maya Polity. In *Classic Maya Political History*, ed. T. Patrick Culbert, School of American Research Advanced

Harris, T. M. G. R. Lock

1990. "The diffusion of new technology: a perspective on the adoption of geographc information systems within UK archaeology." . In *Interpreting Space: GIS and archaeology*, ed. Kathleen M.S. Allen, Stanton W. Green and Ezra B.W. Zubrow, Taylor and Francis, pp. 33-53

Hassig, R.

1991. "Roads, routes, and ties that bind" in *Ancient Road Networks and Settlement Hierarchies in the New World* Charles D. Trombold ed. Cambridge University Press

Haviland, W. A.

1968. Ancient Lowland Maya Social Organization. In *Archaeological Studies in Middle America*. Middle American Research Institute, Publication 26,

1969. A New Population Estimate for Tikal, Guatemala. *American Antiquity*, Vol.34, pp. 429-433.

1970. Tikal, Guatemala, and Mesoamerican Urbanism. *World Archaeology*, Vol.2, pp.186-198.

Hill, J.N.

1966. A Prehistoric Community in Eastern Arizona. *Southwestern Journal of Anthropology*, Vol.22, pp. 9-30.

Hodder, I.

1974 .The *Distribution of Savernake Ware.* Wiltshire Archaeological and Natural History Magazine.

1977. "Some new directions in the spatial analysis of archaeological data at the regional scale (macro)". In *Spatial Archaeology*, ed. David L. Clarke. Academic Press, pp.223-352.

Hodder, I. and C. Orton

1976. Spatial *Analysis in Archaeology*. Cambridge University Press

Hodder, I. R. and M. Hassall

1971. The Non-Random Spacing of Romano-British Walled Towns. *Man*, 6, pp.391-407.

Houston, S., H. Escobedo, P. Hardin, R. Terry, D. Webster, M. Child, C. Golden, K. Emery, and D. Stuart

1998. Investigations *at Piedras Negras, Guatemala 1998 Field Season.* A Report to the Foundation for the Advancement of Mesoamerican Studies, Inc.

Inomata T. and K. Aoyama

1996. Central-Place Analyses in the La Entrada Reagion, Honduras: Implications for Understanding the Classic Maya Political and Economic Systems. *Latin American Antiquity*, Vol.7, No.4, 291-312.

Jackson, J. M.

1990. Building a historic settlement database in GIS. In *Interpreting Space: GIS and Archaeology*, Taylor & Francis, pp. 274-283.

Jones K. T. and D.B. Madsen

1989. Calculating Cost of Resource Transportation: A Great Basin Example. *Current Anthropology*, Vol.30, No.4, 529-533.

Kellie, A. C. D. Orth and K. C. Carstens

1996. Digital Terrain Modeling in Archaeological Mapping. Unpublished ms.

Kohler, T. A. and E. Carr

1996. "Swarm-based modelling of prehistoric settlement systems in southeastern North America. In *Archaeological Applications of GIS: Proceedings of Colloquium II, UISPP XIIIth Congress, Forli, Italy, September, 1996*, (ed.) Ian Johnson & MacLaren North, CD produced by Sidney University Archaeological Methods Series 5, 1997.

Kurjack, E.B.

1974. Prehistoric *Maya Community and Social Organization.* Middle American Research Institute, Publication 38, Tulane University.

Kurjack, E.B. and S. Garza T.

1981. Pre-Columbian Community Form and Distribution in the Northern Maya Area. In *Lowland Maya Settlement Patterns*, ed. Wendy Ashmore, A School of American Research Book,

Kvamme, K. L.

1989. Geographic Information Systems in regional archaeological research and management. In Schiffer M.B. ed. *Archaeological Method and Theory*, Vol. 1, University of Arizona Press, Tucson, pp.139-203.
1990. GIS algorithms and their effect on regional archaeological analysis. In *Interpreting Space: GIS and Arhaeology*, Kathleen M. S. Allen, Stanton W. Green and Ezra B. W. Zubrow (eds.), Taylor and Francis, pp. 112-126.

Landa, D.
1978. *Relación de las Cosas de Yucatán.* Editorial Porrúa S.A. México.

Las Casas, B.
1967. Apologética *Historia Sumaria.* Instituto de Investigaciones Históricas, UNAM, México.

Leach, E. R.

1973. "Concluding Address" in *The Explanation of Culture Change, Models in Prehistory.*

Colin Renfrew (ed.), Duckworth, pp.761-771

Limp, W. F.

1987. The integration of archaeological site patterning through integration of remote sensing , geographic information systems and exploratory data analysis. *In Proceedings of U.S. Army Corps Engineers Sixth Remote Sensing Symposium*, pp.232-262.

Longacre, W.A.

1966. Changing Patterns of Social Integration: A Prehistoric Example from the American Southwest. *American Anthropologist*, Vol.68, pp.94-102.

Losch, A.

1954. The *Economics of Location.* Yale University Press

Lowe, J.W.

1985. The *Dynamics of Apocalypse*. University of New Mexico Press, Albuquerque.

MacEachren A. M.

1980. Travel Time as the Basis of Cognitive Distance. *Professional Geographer*, Vol.32, No.1, 30-36.

Madry, S.L.H. and C.L. Crumley

1990. An application of remote sensing and GIS in a regional archaeological settlement pattern analysis: The Arroux River valley, Burgundry, France. In *Interpreting Space: GIS and Arhaeology*, Kathleen M. S. Allen, Stanton W. Green and Ezra B. W. Zubrow (eds.), Taylor and Francis, pp. 364-386.

Marozas, B.A. and J.A. Zack

1990. GIS and archaeological site location. In *Interpreting Space: GIS and Arhaeology*, Kathleen M. S. Allen, Stanton W. Green and Ezra B. W. Zubrow (eds.), Taylor and Francis, pp. 165-172.

Maler T.

1901-03. Researches *in the Central Portion of the Usumatsintla Valley.*Memoirs of the Peabody Museum of American Archaeology and Ethnology, Harvard University,

Marble D. F.

1990. The potential methodological impact of geographic information systems on the social sciences. In *Interpreting Space: GIS and archaeology*, ed. Kathleen M.S. Allen, Stanton W. Green and Ezra B.W. Zubrow, Taylor and Francis 1996 The Human Effort Involved in Movement Over Natural Terrain. A Working Bibliography. Unpublished Manuscript submitted as Part I of the Final Report to the National Park Service

Marcus, J.

1973. Territorial Organization of Lowland Classic Maya. *Science* Vol.180, pp. 911-916.

1976. Emblem *and State in Classic Maya Lowlands.* Dumbarton Oaks, Washington D.C.

Martin, S. and N. Grube

1995. Maya Superstates. In *Archaeology*, Vol.48 No.6, November/December, pp.41-46

Mathews, P. L.

1975. The lintels of Structure 12, Yaxchilan, Chiapas. Paper presented at the Annual Conference of the Northeastern Anthropological Association, Wesleyan University.

1980. Notes on the dynastic sequence of Bonmapak, part I. *Third Palenque Round Table* 1978, Merle Greene Robertson (ed.), vol.2, pp.60-73.

1985. Early Classic monuments and inscriptions. *In A Consideration of the Early Classic Period in the Maya Lowlands*, eds. Gordon R. Willey and Peter Mathews, Institute for Mesoamerican Studies, Publication 10, Albany: State University, New York, pp. 5-55.

1986. Classic Maya site interaction and political geography. Paper presented at Maya Art and Civilization: The New Dynamics.

1988. The *Sculptures of Yaxchilan*. Ph.D. dissertation, Yale University

1991. Classic Maya Emblem Glyphs. In *Classic Maya Political History*, ed. T. Patrick Culbert, School of American Research Advanced

1995. War In the Western Maya Lowlands. Paper presented in the *1995 Mesa Redonda de Palenque*, Palenque, Chiapas, Mexico.

Mathews, P. L. and L. Schele

1974. Lords of Palenque- the glyphic evidence. In *Primera Mesa Redonda de Palenque*, ed. Merle Greene Robertson, Part I, pp.63-75.

Mathews, P. L. and M. Aliphat Fernandez

1997. Informe Técnico de la Segunda Temporada deCampo del Proyecto Arqueológico El Cayo. Archivo Técnico, Consejo de Arqueología, INAH, Mexico

Maudslay, A. P.

1889-1915. Archaeology. In *BiologiaCentrali-Americana*, London.

Miller, M. E.

1997. "Imaging Maya Art" In *Archaeology*, Vol.50 No. 3 May/June 1997, pp.34-40.

Meave del Castillo, J.

1990. Estructura *y Composicion de la Selva Alta Perennifolia de los Alrededores de Bonampak* Coleccion Cientifica INAH-SEP, Mexico

Morley, S.G., G.W. Brainerd, and R.J. Sharer

1983. The *Ancient Maya*. Fourth Edition. Stanford University Press.

Palij M.

1984. On *the Varieties of Spatial Knowledge: Cognitive Maps and Imagined Terrains.*

Ph.D. Dissertation, Department of Psychology, State University of New York at Stony Brook.

Palka, J. W.

1996. Sociopolitical implications of a new emblem glyph and place name in Classic Maya inscriptions. In *Latin American Antiquity*, vol.7 no.3, pp.211-227.

Pandolf K. B., B. Givoni, and R. F. Goldman

1977. Predicting Energy Expenditure With Loads While Standing or Walking Very Slowly.

Journal of Applied Physiology, Vol.43, 577-581.

Phillips, P. , J. A. Ford and J. B. Griffin

1951. Archaeological *Survey in the Lower Mississippi Aluvial Valley, 1940-1947.* Paper of the Peabody Museum of Archaeology and Ethnology, Harvard University, 25.

Pope K. O. and B. H. Dahlin

1989. Ancient Maya Wetland Agriculture: New Insights from Ecological and Remote Sensing Research. *Journal of Field Archaeology*, Vol. 16, 87-106.

Puleston, D.E.

1974. Intersite Areas in the Vicinity of Tikal and Uaxactun. In *Mesoamerican Archaeology, New Approaches.* ed. Norman Hammond, University of Texas Press.

Redfield, R. and A. Villa Rojas

1934. Chan *Kom a Maya Village.* Carnegie Institution of Washington, Publication 448, Washington D.C.

Reilly, W.J.

1931. The *Law of Retail Gravitation.* New York

Ricketson, O.G., Jr. and E.B. Ricketson

1937. Uaxactun, *Guatemala, Group E,, 1926-31.* Carnegie Institution of Washington, Publication 403,Contribution 1, Washington D.C.

Sanders, W.T.

1962. Cultural Ecology of the Maya Lowlands, Part I. *Estudios de Cultura Maya*, Vol.2, pp.79-121.

1981. Classic Maya Settlement Patterns and Ethnographic Analogy. In *Lowland Maya Settlement Patterns*, ed. Wendy Ashmore, A School of American Research Book.

Savage, S. H.

1990. GIS in Archaeological research. In interpreting Space: *GIS and archaeology*, ed. Kathleen M.S. Allen, Stanton W. Green and Ezra B.W. Zubrow, Taylor and Francis

Schele, L.

1976. Accession Iconography of Chan-Bahlum in the Group of the Cross at Palenque. In *the Art,Iconography and Dynastic History of Palenque*, Part 3, ed. Merle Greene Roberson,
Proceedings of the Segunda Mesa Redonda de Palenque, pp.9-34.
1986. A comparative study of the history of Katun 13 as recorded in the Maya inscriptions. Paper prepared for Elite Interaction in Classic Maya Civilization, School of American Research,
Advanced Seminars, Santa Fe.
1991. An epigraphic history of the Western Maya region. In *Classic Maya Political History: Hieroglyphic and Archaeological Evidence*, ed. T. Patrick Culbert, Cambridge
University Press, pp. 72-101.

Schele, L. , P. Mathews and F. Lounsbury

1977. Parentage *expressions in Classic Maya Inscriptions.* Unpublished ms.

Schele, L. and D. Freidel

1990. A *Forest of Kings: the untold story of the ancient Maya.* William Morrow, New York.

Schele, L. and Grube N.

1994. Tlaloc Venus Warfare, The Peten Wars 8.17.0.0.0 - 9.15.13.0.0. In *Notebook for the XVIIIth Maya Hieroglyphic Workshop*, The University of Texas at Austin.

1995. The last two hundred years of Classic Maya history, Transmission, Termination, Transformation. In *Notebook for the XIXth Maya Hieroglyphic Workshop at Texas*, University of Texas at Austin.

Schele, L. and M.E. Miller

1986. The *Blood of Kings: Dynasty and Ritual in Maya Art.* Fort Worth: Kimbell Art Museum, George Brazillier and Co.

Schele, L. Mathews P.

1991. Royal visits and other intersite relationships among the Classic Maya. In *Classic Maya Political History: Hieroglyphic and Archaeological Evidence*, T. Patrick Culbert (ed.),
Cambridge University Press, pp. 72-101.

Schneider, K. and Robbins P.

1998. GIS and Mountain Environments. *Explorations in Geographic Information Systems Technology*, Vol. 5 UNITAR, Geneva Switzerland.

Sharer, R. S.
1991. Diversity and Continuity in Maya Civilization: Quirigua as a case study. In *Classic Maya Political History*. T. Patrick Culbert (ed.), Cambridge University Press, pp.180-198.

Sheets P. and Server T.L.

1991. "Prehistoric footpaths in Costa Rica: transportation and communication in a tropical rainforest" in *Ancient Road Networks and Settlement Hierarchies in the New World*
Charles D. Trombold (ed). Cambridge University Press

Soule R.G. and R. F. Goldman

1972. Terrain Coefficients for Energy Cost Prediction. *Journal of Applied Physiology*, Vol.32, No.5, 706-708.

Steward, J. H.

1937. Ecological Aspects of Southwestern Society. In *Anthropos*, 32, pp. 87-104.

Stine, R. S. and T. D. Decker

1990. "Archaeology, Data Integration and GIS." . In *Interpreting Space: GIS and archaeology*, ed. Kathleen M.S. Allen, Stanton W. Green and Ezra B.W. Zubrow, Taylor and Francis, pp.134-140.

Stine, R. S. and D. P. Lanter

1990. "Considerations for archaeology database design" . In *Interpreting Space: GIS and archaeology*, ed. Kathleen M.S. Allen, Stanton W. Green and Ezra B.W. Zubrow, Taylor and Francis, pp. 80-89.

Taylor, P.J.

1975. Distance *Decay in Spatial Interactions.* CATMOG 2, Geo Abstracts Ltd., University of East Anglia, Norwich

Thompson, J.E.

1930. Ethnology *of the Mayas of Southern and Central British Honduras.* Field useum of Natural History, Pub. 274, Anthropological Series, Vol.XVII, No. 2. Chicago.

Tobler, W. and S. Wineburg

1971. A Cappadocian Speculation. *Nature*, Vol.231, No.5297, pp.39-41 (reprint pp.1-6).

Tovalin Ahumada, A.

1996. Arquitectura y Patron de Asentamiento. Paper presneted at the Maya Meetings at Calgary, Archaeology Department, University of Calgary, Calgary.

Tovalin Ahumada, A. and Velazquez de Leon Collins J.A.

1995 .Proyecto Arqueologico Bonampak, Tercera Temporada de Campo 1995 Informe Técnico. Archivo Técnico del Consejo de Arqueología INAH, Mexico.

Trombold, C.D.

1991. "An introduction to the study of ancient roads" in *Ancient Road Networks and Settlement Hierarchies in the New World.* Charles D. Trombold (ed.) Cambridge University Press pp.1-9.

Turner, E.S., N.J. Turner, and R.E.W. Adams

1981. Volumetric Assessment, Rank Ordering, and Maya Civic Centers. In *Lowland Maya Settlement Patterns*, ed. Wendy Ashmore, A School of American Research Book, pp

Vita-Finzi, C. and E. Higgs

1970. "Prehistoric economy in the Mount Carmel area of Palestine: site catchment analysis. In *Proceedings of the Prehistoric Society*, London, pp.36, 1-37.

Vogt, E. Z.

1964. "Some implications of Zinacatan Social structure for the study of the ancient Maya." In *35th International Congress of Americanists 1*, INAH, México, pp. 307-319.

Warren, R.E.

1990a. Predictive Modelling in archaeology: a primer. In *Interpreting Space: GIS and Arhaeology*, Kathleen M. S. Allen, Stanton W. Green and Ezra B. W. Zubrow (eds.), Taylor and Francis, pp. 90-111.
1990b. Predictive modelling of archaeological site location: a case study in the Midwest. In *Interpreting Space: GIS and Arhaeology*, Kathleen M. S. Allen, Stanton W. Green and Ezra B. W. Zubrow (eds.), Taylor and Francis, pp. 201-215.

Waters, N.

1995. "The most beautiful formulae in GIS" in *Spatial Reasoning for Effective GIS*.

Fort Colins, Colorado, pp.175-192.

Williams, I., W.F. Limp and F.L. Briuer

1990. Using geographic information systems and exploratory data analysis for archaeological site classification and analysis. In *Interpreting Space: GIS and Arhaeology*, Kathleen M. S. Allen, Stanton W. Green and Ezra B. W. Zubrow (eds.), Taylor and Francis, pp. 239-273.

Williams, I., W.F. Limp and F.L. Briuer

1990. Using geographic information systems and exploratory data analysis for archaeological site classification and analysis. In *Interpreting Space: GIS and Arhaeology*, Kathleen M. S. Allen, Stanton W. Green and Ezra B. W. Zubrow (eds.), Taylor and Francis, pp. 239-273.

Willey, G. R.

1981. "Maya lowland settlement patterns: a summary review" *in Lowland Maya Settlement Patterns*, ed Wendy Ashmore, University of New Mexico Press, pp. 385-415.

1953. Prehistoric Settlement Patterns in the Viru Valley, Peru. Bureau of American Ethnology, Smithsonian Institution Bulletin No.155, Washington D.C.

1987. Essays *in Maya Archaeology*. University of New Mexico Press, Albuquerque.

Willey G. R., W. R. Bullard Jr., J. B. Glass, and J. C. Gifford

1965. Prehistoric *Maya Settlements in the Belize Valley*.

Papers of the Peabody Museum of Archaeology and Ethnology, Harvard University Vol. LIV.

Willey, G.R. and J. A. Sabloff

1993. A *History of American Archaeology*. Third Edition, W.H. Freeman and Company, New York.

www.ingramcontent.com/pod-product-compliance
Lightning Source LLC
LaVergne TN
LVHW071219210926
848482LV00015B/856